Agile Methods for Safety-Critical Systems

A Primer Using Medical Device Examples

Nancy Van Schooenderwoert and Brian Shoemaker
© Copyright 2018

Contents

Preface	ii
Introduction	v
Chapter 1 – A Brief Introduction to Agile	1
Chapter 2 – Origins of Agile	12
Chapter 3 – Benefits of Agile	17
Chapter 4 – Applying Agile: The Importance of Stories	33
Chapter 5 – Applying Agile: Iterative and Incremental, not Linear	42
Chapter 6 – Applying Agile: Agile Teams and Environments	52
Chapter 7 – Applying Agile: Cumulative Documentation and Risk Management	57
Chapter 8 – So (We've Sold You on the Benefit and) You've Decided to Become Agile	65
Chapter 9 – Agile Planning Techniques	72
Chapter 10 – Tracking Progress and Accelerating Learning	83
Chapter 11 – Scaling Up	97
Chapter 12 – Simple, But Not Easy	105
Afterword: Recommended Reading	109
About the Authors	112
References	114

Preface

(Brian) Quality—which includes safety—is a critical feature in software-driven medical and diagnostic devices. Unfortunately, a question that long bothered me was how we can ensure quality, safety, and performance without having that work viewed merely as an unavoidable cost. I had conducted software quality work for some years, and I always felt that I was an anchor dragging down the engineering group without adding any value. We documented and validated our software because we had to, not because anyone in charge in our company felt it gave any benefit.

The first time I saw the test-first approach that Agile teams were using, a light came on. I knew I finally had my answer. What I saw were quality-control activities that were well integrated into the development process rather than being left until the end—activities that indeed led to better and safer products rather than simply being checkmarks in boxes or stamps of approval after all the "important" work had been completed.

During my in-vitro diagnostics days, our engineering VP, when discussing our instrument software (in particular the user interface), used to grouse: "Don't ask the customers! They don't know what they want!" He was only partially right; clinical laboratory technicians, who were usually overworked from running a wide variety of instruments day in and day out, had a hard time describing to us what features would make an instrument more reliable and less challenging for them to use. After all, they weren't design engineers. What I realized later, though, was that if they could see a prototype in action, they'd quickly let us know whether we had gotten the features right for their workflow. We, the engineers, had to understand them, the users—not the other way around.

Between the test-first approach and the practice of demonstrating early and often, I realized that this "Agile" thing had huge potential. That understanding brought me to collaborate with Nancy Van Schooenderwoert of Lean-Agile Partners. From Nancy, I've learned a great deal more about what Agile is all about. Combining our strengths has led to numerous workshops that we've presented to both medical device and Agile audiences over the past eight years.

However, not everyone can attend a workshop. That, combined with the importance of safety and quality in the medical device sector, was the genesis of this book.

(Nancy) It never made sense to me that bugs are inevitable in software. They couldn't get there unless we put them in, so why can't we just not put them in? My work in flight simulation gave me many opportunities to see how the aerospace industry uses systemic measures to prevent defects and learn from every accident so the right mitigations can be implemented.

Later work in factory automation and medical devices gave me the chance to pick up ingenious techniques from the teams I worked with. In one medical device project, I needed to port software to new, miniature display hardware. Rather than take six to eight weeks to do the port the usual way, I was able to do it in two weeks by using quick tests and swapping the hardware into the loop and back out. Today we'd call that "test-first" or "test-driven" development. That project was the first time I saw (or even heard of) a software project that finished ahead of the *original* schedule—not a schedule that had already been slipped a dozen times. We were a few weeks early, mainly because the technical lead had us doing real unit tests and preserving them to form a regression test suite. This was long before anyone called these practices "Agile."

Soon, my tool bag included the unit test practice, the test-first practice, and a really elegant way to embed tracing into the source code so you could troubleshoot in a lightweight way. But I couldn't convince later contractor teams I worked with to adopt a combined set of these techniques, and the techniques could not work unless everyone on the team would cooperate in their use. I knew each was a time-saver, and I wondered whether the effect would be additive or multiplicative if they were used together.

I had an opportunity to go to a new company that was building a spectrometer instrument, where I'd be able to form a team and use modern software practices. During that project, we cooperated as a real team to use all the practices, and even created more. Those practices definitely had a multiplier effect, and were so powerful we could hardly believe our own data. According to data published by

Capers Jones, our defect density was an order of magnitude lower than the best traditional teams. We could directly compare our data with his because there was enough information on how it was gathered. Capers Jones even looked over our computations, and our results and agreed with that conclusion.

We also never needed a bug database, and our open defects never exceeded two all through the project! We averaged about 1.5 defects a month getting through our unit-test defenses, and that was for the whole team of five (though it varied a little throughout the project).

Since then, other Agile teams have been able to take quality to that level and even higher. I continued coaching teams and their managers to become Agile, and helped them extend their agility to hardware.

I met Brian at a local professional event. He was curious about Agile, and I had some software that I could use to show him how a unit-test suite works. That led eventually to our collaboration to help people in safety-critical work—and especially in medical device work—understand that Agile techniques open up a level of quality and reliability not achievable with older methods. Agile's speed is a welcome side effect. You can't get quality by aiming for speed, but you can get amazing speed by aiming for uncompromisingly high quality.

Introduction

Medical device recalls occur almost every day, and a large fraction of those recalls are attributable to software. Regulatory bodies first recognized that software could fail in unexpected and unsafe ways when the Therac-25, a linear accelerator used for cancer therapy, had a series of accidents that exposed patients to some 200 times the full-body lethal dose of radiation.[1] Since then, a number of other incidents have resulted from software issues; these examples come verbatim from the public FDA medical device recalls database.[2]

- GUI[a] software interface between a diagnostic instrument and a sample handling workstation could assign sample results to the wrong patient.[3]
- A Software Application Card, used with an external programming device to set up parameters on specific implantable infusion pumps, allowed users to enter a periodic bolus interval mistakenly into the minutes field, rather than the hours field. Numerous adverse events occurred, including two patient deaths and seven serious injuries due to drug overdose.[4]
- A dialysis device was recalled after eleven injuries and nine deaths because too much fluid was removed from the patient when the caregiver overrode the device's "incorrect weight change detected" alarm.[5]
- A picture archiving and communications (PACS) system was recently recalled for software issues which could cause (1) potential data loss, (2) study mix-up, (3) incorrect measurements on multi-frame images, (4) dearchiving issue, and (5) unauthorized access of data due to inadequate permissions for shared folders.[6]
- An infusion pump was recalled because if it received even a single byte of data into its serial port (intended only for monitoring the pump) from an external system, the pump would go into a failure condition. If the failure occurred during infusion, the pump would need to be powered down and restarted.[7]

---

[a] Graphical user interface

- A software anomaly in an infusion pump caused an audible alarm and stopped the function of all channels in use, and therefore caused an interruption in therapy.[8]
- A software error in a cranial navigation system (Image Guided Surgery System) could potentially intensify small inaccuracies arising from individual steps of a complex navigation procedure, and cause an inaccurate display of instruments by the navigation system compared to the actual patient anatomy.[9]
- A mobile application for diabetes management contained a program error in the Bolus Advisor feature: when the OS region of the phone setting changed, the unit of measure within the app could unexpectedly change, creating a risk the app might not transfer the blood glucose result or the user might not correctly input numerical values for carbohydrate used for bolus advice.[10]
- A Vision System was recalled after the manufacturer identified both software and hardware problems associated with unexpected system loss of power (shutdowns), unintended system error messages, unresponsive touchscreens, and system setting and infusion performance problems. These events could cause eye injuries, including blindness.[11]

Plenty of systems with embedded software are safety-critical: in aerospace, in transportation, in industrial control. This book focuses on medical devices because that field serves as a useful umbrella to cover aspects of development that are important to safety-critical work.

While safety is obviously critical, at the same time, innovation is the lifeblood of just about every industry—and medical technology is a prime example. A "production" mentality where everything goes according to plan will view variation and experimentation as waste, but in development such flexibility is necessary for creating new solutions and finding ways to make them work. Agile methods thrive in situations where a new pathway needs to be found. Agile is the best core foundation for any company that truly needs to innovate.

Agile methods give us access to levels of quality and safety that are not attainable with traditional linear, inflexible approaches (often called "waterfall"), where development can only proceed in one direction. Rare exceptions exist, of course: for example, NASA achieved amazing software and systems quality levels using traditional methods, but only by use of extremely labor-intensive practices that ordinary industry could never afford. After all, failure is truly not an option in space flight![b] After teams and companies fully adjust to it, the Agile approach makes this level of quality routine and even cheaper than older methods.

The classical maxim is "You can have it good, fast, or cheap: pick any two." Building a high-quality product quickly would always be extremely expensive. Developing a product quickly and inexpensively always meant that it would compromise on quality. And so on. With Agile, the significant advantage is that quality, speed, and cost are no longer enemies. You don't have to trade any one of them for the others.

Agile practices are able to combine the knowledge of a team smoothly, so that the waste and misunderstandings of usual communication by relying solely on written specifications is eliminated. Teams can respond in real-time much as if they were one organism.

Agile helps teams in other ways as well. Studies show that employee engagement at work is at an all-time low. However, Agile practices, when done properly, make for a very cooperative and stimulating workplace culture that employees love. Menlo Innovations is a great example of this Agile effect. Everyone is co-located and in touch with what's expected of them, because all the work assigned to each person is visible on a wallboard display. Each employee knows that they're doing something a client cares about, because they have a feedback connection to customers. Simple, visible mechanisms like the wallboard display mean less need for meetings to talk about status and more time for the real work.

---

[b] Even NASA recognized the need for incremental planning when many unknowns are present —the space shuttle and Apollo 8 through Apollo 11 were built iteratively.

The type of culture that Agile fosters helps people to see and appreciate more clearly everyone's contribution to the whole. This is a great foundation to improve issues related to sexism or racism, which are otherwise very hard to deal with.

The cross-function nature of Agile teams mirrors the makeup of the product itself as it evolves through short iterations, or "sprints," where its correct behavior is demonstrated. Team members thus have the opportunity to broaden their skill sets. Involvement in a focused core group also supports team members' need to learn and grow in their own discipline through Communities of Practice – peer learning groups that cover a functional area such as software design or test automation.

In short, Agile is the answer to the safety, quality, and innovation needs of every safety-critical industry. We've put this book together to show you why—and where you should start as you move to an Agile approach.

You, our reader, will find this book most meaningful if you have some knowledge of software development and development models, but you don't need to be a trained programmer. You also don't need detailed knowledge of Agile, though we expect you've heard of it. We give a quick overview of what it is, but we don't describe it in detail.

We caution you that this book is a primer, not a complete how-to guide. Knowing *about* a thing is not the same as knowing the thing itself. You can read all you want about learning German or about learning to play the piano, but none of that will give you the fluency to read Goethe and converse with Aunt Gertrude in Stuttgart, or make you competent to play a Mozart sonata. This book is not a "Swiss Army knife" – it can't possibly equip you to handle every situation that will come up in your development process. Rather, this book has the *about* knowledge that we have found valuable to those embarking on their Agile journey. Our intention in this primer is to show you enough about applying Agile, planning projects and products, tracking work and measuring team accomplishments, and avoiding stumbling blocks so that you can engage a coach with confidence. To help you really *know* the Agile approach, we believe coaching is best. And coaching begins with a little training, not a lot.

Chapter 1 – A Brief Introduction to Agile

The first step in this journey is getting to know a bit more about Agile. You've likely heard of Agile before, possibly as some offbeat, undisciplined approach that website developers use to slap something together in a hurry. We understand how suspicious it may seem if developers cover much more ground than they did before and with higher quality – traditional wisdom says that they must be taking shortcuts. We beg to differ: effectively employed, Agile is the safest approach we know for tackling new product work.

Some product work is based on known techniques that you've used before. Other work is new—possibly new to the world or maybe just something that you haven't previously done in-house. If no part of the product or its construction work is new, then we are by definition building a commodity product, which requires manufacturing effort but no development effort. A helpful metaphor is the "Cook or Chef" idea proposed by the Poppendiecks,[1,2] which is illustrated below. Agile methods are best suited for the new portion of the work, while Lean practices, which rightly revolutionized manufacturing, are already well respected for the known part of the work. Agile and Lean are built on principles that are highly compatible.

Well-structured problems *Poorly-structured problems*

Manufacturing   **Development**

Production *Cook*   **Design** *Chef*

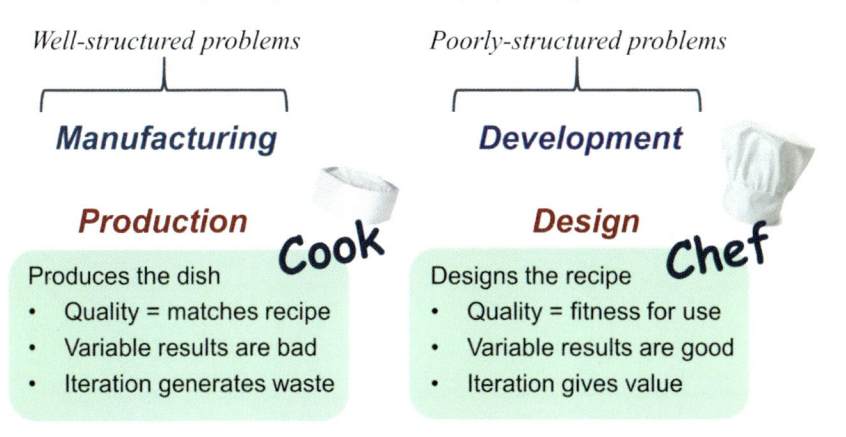

Produces the dish
- Quality = matches recipe
- Variable results are bad
- Iteration generates waste

Designs the recipe
- Quality = fitness for use
- Variable results are good
- Iteration gives value

Figure 1 Contrasting Production "Cook" vs. Development "Chef"

We'll go into more detail later, but here are some crucial points:
- Proper Agile is a discipline that teams actually follow.

- Well-functioning Agile is akin to a well-functioning quality system, such as 21 CFR Part 820 (the Quality System regulation) or ISO 13485.[13] It establishes a context of quality and safety within which all other activities take place. Neither the Agile approach nor the quality system is a fixed checklist.
- Agile applies to many areas, not just to software development. Methods akin to Agile have been used in mechanical and electrical engineering for years. In addition, the flexibility and adaptability of Agile apply to areas beyond engineering, including quality assurance, regulatory affairs, marketing, and, of course, management. A truly Agile approach demands that all these functions communicate regularly.
- Agile is a mindset, not a cookbook method that can be imposed from above and followed by rote or written up in a standard operating procedure.

Agile *per se* was articulated in 2001 as an explicit shared set of values and way of working. The Agile Manifesto expresses the most abstract of three levels for that mindset. Twelve Principles then lay out the approaches through which that mindset is implemented; note that flexibility and adaptability are essential elements (box next page).[14] Although the word "software" is used, many companies have applied these ideas to whole products and business processes.

Agile Methods for Safety-Critical Systems

> **Manifesto for Agile Software Development**
>
> We are uncovering better ways of developing software by doing it and helping others do it. Through this work we have come to value:
> - Individuals and interactions over processes and tools
> - Working software over comprehensive documentation
> - Customer collaboration over contract negotiation
> - Responding to change over following a plan
>
> That is, while there is value in the items on the right, we value the items on the left more.
>
> **Principles behind the Agile Manifesto**
>
> We follow these principles:
> - Our highest priority is to satisfy the customer through early and continuous delivery of valuable software.
> - Welcome changing requirements, even late in development. Agile processes harness change for the customer's competitive advantage.
> - Deliver working software frequently, from a couple of weeks to a couple of months, with a preference to the shorter timescale.
> - Business people and developers must work together daily throughout the project.
> - Build projects around motivated individuals. Give them the environment and support they need, and trust them to get the job done.
> - The most efficient and effective method of conveying information to and within a development team is face-to-face conversation.
> - Working software is the primary measure of progress.
> - Agile processes promote sustainable development. The sponsors, developers, and users should be able to maintain a constant pace indefinitely.
> - Continuous attention to technical excellence and good design enhances agility.
> - Simplicity—the art of maximizing the amount of work not done—is essential.
> - The best architectures, requirements, and designs emerge from self-organizing teams.
> - At regular intervals, the team reflects on how to become more effective, then tunes and adjusts its behavior accordingly.

At the visible level, a large number of technical practices are available for Agile teams to use in their work. Which practices a team uses are up to them. Ahmed Sidky of IC Agile summarized the way Agile values and principles support the Agile Practices in this diagram (Figure 2):[15]

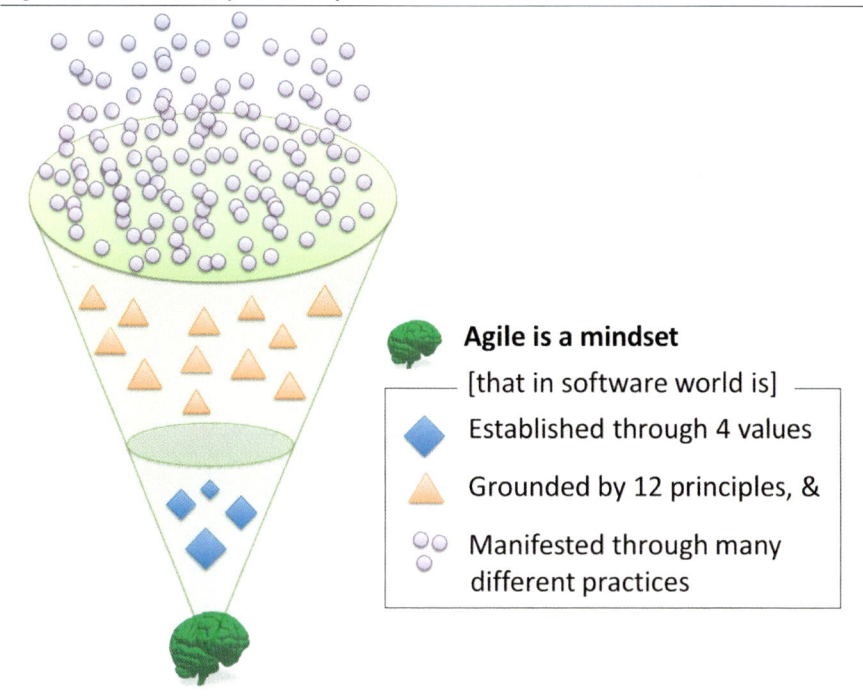

Figure 2 Three levels of the Agile mindset

Although flavors of Agile (e.g., extreme programming or Scrum) focus on specific sets of practices, it's absolutely fine to mix those practices based on what a team finds works for them and achieves the desired outcomes.

Agile Control Loops

Agile approaches can be viewed as two control loops. In engineering, a control loop is a series of operations that uses feedback to control something. A thermostat is a simple example: it controls room temperature by sensing the current temperature (as feedback) and deciding whether to keep the heater on or turn it off.

Agile works through similar feedback loops. An Agile process for building robust software will sense several things about that software: whether it's passing all the unit tests, whether sufficient test coverage is present, whether the software has duplication or close coupling, and any other signs of poor condition. If these are present, a signal is triggered so the problem can be addressed. That signal happens via the team's retrospective (the Agile term for regular meetings designed to review and learn from the work just completed and to guide the work going forward).

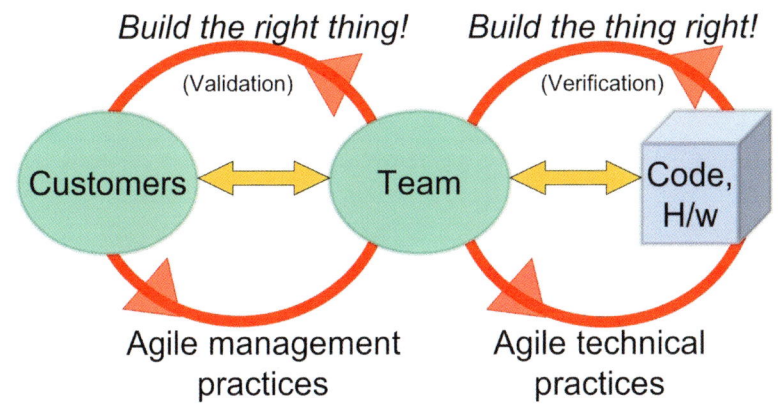

Figure 3 The Dual Feedback Loops in Agile

An Agile process for building the right software will sense what the end users, maintainers, and field servicers (the customer community in general) find valuable, collecting this information and organizing it into a form that facilitates building the product incrementally, with fully working features in place as early as possible each step of the way. That form is what we call a User Story. If there is trouble with this control loop, it should trigger a signal to the wider organization, beginning at the team's retrospective.

"Live" Control Mechanism for What to Build and How to Build It

Similar to the feedback loops, Agile provides practices and controls that help you create useful products that are also built well. It makes sense to separately control the processes for "building the right thing" and for "building the thing right." Both are important, and good performance in one cannot adequately compensate for poor performance in the other. Agile methods give clear, direct ways to control and monitor each of these two key processes.

For example, the Product Owner role calls for one person[c] to have the responsibility and authority to decide what functionality to build and the order in which to build it. Activities like the "sprint review" and "sprint retrospective" occur every two to four weeks, at intervals typically termed "sprints" or "iterations," and are explicitly held to acquire new information and adapt the work to this information.

For the "build it right" side, the Agile community has created practices like "test-driven development" and "continuous integration," along with tools such as unit test frameworks for all of the programming languages in common use. Together, these practices enable a team to continually answer the questions "Are we building this right? Do all the tests run cleanly?"

By placing a "live" control loop around each of these two activities, the Agile approach allows you to plan what is plannable at the start, and to adapt to emerging changes as your project proceeds.

Test as You Go

In software work, finding and fixing defects takes up a large percentage of time for software teams—30 percent to 50 percent is common. Part of the problem is the way troubleshooting effort grows: there's a multiplier effect. If code has two bugs, the repair is much more difficult than simply twice the effort for one bug. And if three bugs are potentially interacting, the problem scales even further. When we can fight just one bug at a time, the effort per bug is reduced just as drastically.

This effect was reported by Nancy's early Agile team on the spectrometer project described in the Preface. The software team had an open defect list that never went above two defects for the duration of a challenging embedded systems project. The team spent negligible time on defect find-and-fix activities because they used incremental testing techniques and automated testing; this prevented most defects, so they rarely had multiple defects interacting. The net

---

[c] The Scrum framework says the Product Owner role must be one person, but I (Nancy) have found a Product Owner Team to work well in certain cases.

result was that practically 100 percent of the team's capacity was available to the business for new feature development all through the project. For most industry software teams, this effect alone would double their productivity.

Here's another example. Neil Johnson (author of the blog "Agile SoC") created a test runner for programmable hardware and used that to shift his SoC (system on a chip) verification work from traditional to test-driven. He benchmarked the change in the number of defects he was able to catch and fix in his SoC coding work. He says that this change has produced a fifteen-to-one reduction in defects for his own work.

Let's go back to that list of recalls mentioned in the Introduction. Software-based defects are by far the fastest growing reason for FDA recalls since 2007. Between January 1, 2007 and December 31, 2013, there were 1678 computer-related medical device failures affecting eighteen million medical devices. Of those, 64 percent were software-related. Over that same period, software-related failures as a percentage of all failures more than doubled.[16]

According to a VersionOne 2014 survey, organizations adopting Agile methods saw a 78 percent increase in their software quality. Using the FDA recall numbers, that means that better quality based on Agile development could potentially eliminate over 800 of those recalls.[17]

Teams Should be Cross-Functional

Another value of Agile comes from cross-functional teams. Agile software teams include developers, testers, and other specialized people (such as database administrators, algorithm specialists, or web application specialists) depending on the type of application. There is no conflict with the regulatory need for independence of review when testers work alongside developers. The FDA Quality System Regulation (21 CFR Part 820) states:

> *Each manufacturer shall establish the appropriate responsibility, authority, and interrelation of all personnel who manage, perform, and assess work affecting quality, and provide the independence and authority necessary to perform these tasks.*

The tester who collaborates with developers on a product team is still independent; he/she reports to a quality manager and is responsible for quality aspects of the emerging product.

Likewise, there is nothing in the Agile principles that prevents their application to firmware and hardware aspects of medical device development and operations. Simulation and quick-turn suppliers make it increasingly possible to keep working hardware available on the same cadence with software iterations.

Using separate or joint hardware and software teams is a choice to be made based on your situation. Either way can be valid from the Agile perspective. Both the FDA and European regulatory standards expect you to define your process and the controls you are using to guide it; they are silent on your choice of process.

Agile Helps You Avoid "Unrecoverable Losses"

Once lost, some things cannot be recovered:

- Time – The days and weeks lost to needless debugging and feature-bloat can never be replaced
- Revenue – Wasted time translates to wasted revenue in knowledge work
- Market share – While market share can in theory be recovered, as a practical matter it is very difficult and expensive to recover it once it's been lost

Agile teams create solid foundations and build upon them very pragmatically. Regression testing is automated so that testing can keep pace with the growing code base. Changes and new features are specified in the test framework, and the code is "grown" within that framework until it passes all the tests. In addition, Agile incorporates regular customer reviews of incremental releases, and these put a bound on how far off track a developing product or service can go before getting a needed course correction.

In essence, the steady progress of constructing tests and automating them replaces the hit-or-miss nature of old-style debugging efforts. The result is a steadier and more predictable pace of work.

Agile Isn't One Size Fits All

The Agile approach encompasses technical and management practices that are designed to handle uncertainty while delivering fast time-to-market with far fewer defects than prior development processes. However, uncertainty is by its very nature impossible to plan for in detail. Agile practices similarly cannot be adopted in a cut-and-paste or checklist sort of way; there is an art to it. Teams need the right amount of autonomy. Requirements and project components vary. Emergent changes are inherent in Agile development processes. In addition, it takes time and practice to become proficient at the all-important skill of breaking the work stream down into the smaller elements (the user stories) that are necessary for a smooth delivery cadence. Agile business and technical practices with a flow of right-sized stories are the building blocks for alignment between business, engineering, and customers. That alignment and the benefits it brings are the goals of Agile methods.

Each of the various Agile methodologies (e.g., Scrum, Kanban, XP, Crystal, and others) have its sweet spot. Each has delivered real business success in some situations. Bringing Agile methods that last mile from theory to actual practice in *your* company requires the judgment that comes from a deep understanding of the Agile principles given in the Manifesto combined with a knowledge of your company's circumstances and the true constraints at each step along the way.

Agile Is Not an SOP!

(Brian) I've been a quality assurance manager in clinical-trial data management or in-vitro diagnostics for a number of years, so I'm not surprised when a QA manager new to the Agile idea asks me what the standard operating procedure for Agile should be. I was dismayed, however, when I started working with a diagnostics client recently, and the first request from the software development manager was for my help in framing an SOP for Agile.

Very little I could say in that meeting seemed to convince him that an "Agile SOP" wasn't what he needed. SOPs should, in general, outline what needs to be generated—not how to carry out the steps of development. Agile, I tried to explain, is a framework for being ready to adjust to change, willing to adapt and modify methods, and poised to improve how and what one is developing. An Agile SOP would defeat the very purpose of using Agile.[d]

Our Recommendation for Rapidly Gaining Proficiency with Agile Methods

Organizations are accustomed to using training courses when it's time to upgrade skills. Coaching is necessary in addition to training if you intend to gain real proficiency for your Agile pilot teams within twelve to eighteen months. But coaching is a far less common service for organizations to buy. They don't know what to expect, or how to tell if it's working well enough. The simple fact is, applying Agile concepts is highly dependent on the context: the company culture, the type of business, the product, the market, and the people. In this way it's different from other skills you might use training to gain, such as learning a new graphics application or learning accounting, for example. In cases like those, it's just you and the subject matter, so a training course and a reference book will do the job. Not so when learning to use Agile methods. The Agile techniques that you'll need vary far too much depending on your situation.

Learning on your own is, of course, an option. It can work, but it's not rapid. Often it leads to frustration and giving up, or to very modest improvement.

---

[d] Of course, your software development SOP needs to specify your software development life cycle.

Since we're talking about *rapidly* gaining proficiency with Agile methods, here is what we recommend: Start with some training—not only for the project team, but for managers too. Agile needs support from managers to deliver its best improvements. Get coaching to help you implement Agile practices in both the technical and management spheres. Interview several possible coaches. Let the team have a say in which coach is chosen. Ask around. Find out what sort of experience others have had with coaching. Check the coach's references.

Understand the most important problem that you want to solve, and ask the potential coach how they will help you do that.

Look for real experience doing coaching, rather than merely lots of certifications. There are many coaches who mainly teach training courses, and rarely coach teams or managers. Some Agile coaches work internally as regular employees, and others are independent. A good pattern is to begin with an independent coach and have them help you mentor new coaches from among your employees to help spread the Agile practices in a cost-effective way.

Of course, you want to get your people proficient in Agile methods, not create an unending dependence on the coach! Discuss this goal early with the prospective coach. There is no substitute for doing your homework. Getting coaches through an agency, betting on certificates, or hiring the lowest-priced one are risky, well-travelled paths. Do your homework!

Remember the difference between knowing about something and truly knowing it. A child may be fluent in Spanish but not know how to describe its grammar. That's knowing it. A linguist may know all about the language's structure but not be able to speak it fluently—that's knowing about. Both kinds of knowledge are valuable. Coaching gets you fluency. Training gets you ready.

The rest of this book contains the "about" knowledge that we believe will be useful whatever learning path you take.

Chapter 2 – Origins of Agile

Agile Arose from Hardware Practices

(Nancy) The earliest Agile software practices emerged directly out of work with hardware! I can give two examples: my own work and that of the first person I met whose team had code as bug-free as that of my own team.

Having been an electronics designer, taking circuit boards through the stages of breadboard, wire wrap, and so on seemed to tie in with the ideas I had read about using a spiral lifecycle for software development. I worked with a group of electronics engineers at Singer-Link Flight Systems, where we designed flight instrument circuitry for professional flight simulators (used to train pilots and astronauts). We also wrote the software to drive that circuitry. Hardware work was naturally done in iterations; we had to test thoroughly at intervals to cut the risk of defects. As I moved to creating more software, that pattern still made sense to me. Years later, I was technical lead for an embedded software team where we borrowed ideas like spiral lifecycle, modularity, checklists and so on from hardware.

After the Agile Manifesto came out in 2001, I met Ward Cunningham at one of the early Agile conferences. His team was seeing defect rates as insanely low as mine! Ward had taught test-first programming to Kent Beck, who went on to write *Extreme Programming Explained*.[18] It was that book which had shown me there were others using most of the same iterative software development techniques that I had my own team using in 1998–1999. I didn't know until I met Ward, though, that others were also getting such low defect rates – on the order of a few per month for a team of five or six.

Ward's research work at Tektronix between 1977 and 1987 in testing the early microprocessors was the forerunner of TDD (test-driven development), a key Agile software practice. In a 2004 conversation, I asked Ward how his background in embedded software influenced the practices that later became extreme programming. He replied:

> *"When I left the university and started working, I spent ten years in a research lab, and one of the items we researched was microprocessor debugging. Microprocessors were new in that era, and I came to a conclusion that you had to get the machines to tell you what they were doing. And that if you wanted to trust what you got the machines to tell you, that part had to be very simple... That aspect was imprinted on everything I've done since."[19]*

There may be others who, unknown to us, followed a similar path starting from hardware work. Or others who started from purely software work and came to Agile. My point is that hardware and Agile are not at all incompatible!

Hardware is Often Mission/Safety-Critical

Products that have both hardware and software, such as space flight, aviation, defense systems, transportation, navigation, machinery, and medical devices/instrumentation are often safety-critical systems, or at least mission-critical. When you're building systems like these where failure is just *not* an option, you need ways to find and fix problems as soon as they develop. For that you need good testing to detect problems and current, accurate documentation to understand what you are seeing.

Leaving test and documentation until the end of the project, or until the end of long phases like three or six months, is actually a very high-risk approach to systems development. Interfaces between modules become difficult to debug. User interfaces become thorny to rework. Safety risks discovered later on are much more painful to mitigate. Engineers begin to forget how specific features were implemented.

A common objection to using Agile for hardware is that hardware lead times are too long to work with the Agile cadence. However, engineers have been developing mockups, breadboards, and prototypes for many years, without calling the process Agile. Those practices are all mitigations for the fact that the physics of hardware development is indeed different: long lead times and physical wear are a fact of life. Simulation is also a mitigation that can be indispensable, and it's not limited to systems involving hardware.

It's not necessary to build new hardware or new hardware features in two- or three-week periods for hardware development to proceed very well alongside Agile software work. What matters is that there be a working version of the hardware ready to be used at each interval when new software is ready.

Lean Principles expressed via different practices in other contexts

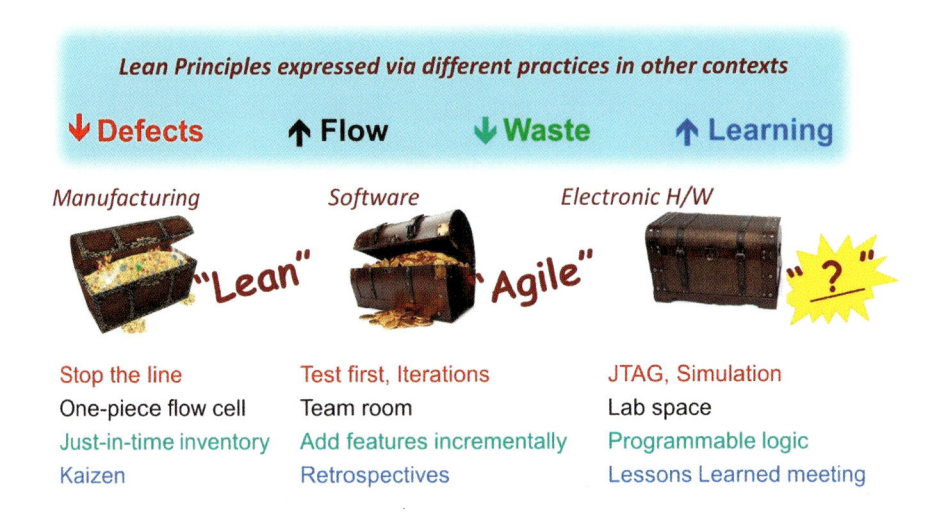

↓ Defects   ↑ Flow   ↓ Waste   ↑ Learning

Manufacturing — "Lean"
Software — "Agile"
Electronic H/W — "?"

Manufacturing	Software	Electronic H/W
Stop the line	Test first, Iterations	JTAG, Simulation
One-piece flow cell	Team room	Lab space
Just-in-time inventory	Add features incrementally	Programmable logic
Kaizen	Retrospectives	Lessons Learned meeting

Figure 4 How Lean principles are expressed in different disciplines

Another example is when hardware and software cooperate to sense an over-temperature condition or to run a power-on self-test. The interaction sometimes can be scripted in advance, but experienced designers know that there is a fair amount of discovery work needed to understand how components will degrade over time, or how their performance will vary when the environment includes vibration, heat, cold, electrical noise, and so on. This type of co-evolution across electronics, software, mechanical components, and materials is nothing new to the engineering world.

Principles Stay Constant; Practices Vary

In Figure 4, the principle of lowering defects is expressed through the well-known "stop the line" practice in lean manufacturing, where any worker seeing defective parts being replicated can pull a cord and stop the production line. That same idea of keeping defects low is expressed in software by writing a small piece of code to test a portion of your overall production code before you even write that production code. This reduces defects by getting you to think clearly about what testable code behavior you need to generate. This process substitutes for the physical line stopped in manufacturing. And as discussed, the same principle of lowering defects in electronic hardware can be expressed through the use of JTAG technology.

Other principles are also expressed through different technical practices in different disciplines:
- Increasing the flow of work (more throughput for same effort)
- Decreasing waste
- Increasing learning so the team can keep improving their process

For manufacturing and for software, the sets of practices have names: Lean and Agile. There is no special name for the equivalent technical practices in electronics, or many other disciplines.[20]

Advent of Computing Power

You could argue that Agile itself is not new either. The individual practices used by Agile teams were already known before the term was coined. Some of them had been around for years. So what was the catalyst for bringing these ideas together under the heading "Agile?" An important factor was that by the late 1990s, software teams began to have easy access to computing power that could support the kind of near-continuous testing needed for fast iterations. Without that, the effort to integrate and test software was just too large to do frequently. It was commonly done at three- or six-month intervals. The ability to integrate each software change *daily* into the main trunk created transparency in the growing codebase that teams never had before. The mere presence of that view into the code had a profound effect. It helped to grow a learning culture within software teams.

Software as a Complexity Magnet

In embedded systems work, it's common to move functionality from hardware to software when you have the choice. Why hardwire the control of an indicator light when you can control it from software and then later decide whether it should blink fast or slowly, or even change what it's indicating? Software has infinite flexibility — up to the point where we make it so complex that we inject new bugs every time we try to touch it! But if we're using Agile technical practices to keep the codebase clean and well structured, then software remains very flexible. However, over the years many other kinds of complexity have migrated into software. In addition to functionality that was once implemented through electronics, a variety of mechanical controls are now through software. For example, car brakes are not mechanically actuated, but routed through software in the dozens of microprocessors modern cars contain. In banks, procedures that were once on paper in policy documents are now coded into databases as the "business rules."

Throughout modern organizations and products, software has become a complexity magnet because of its tremendous flexibility. That is why a new approach was first needed in software development. Traditional engineering development methods broke down, and we were forced to find a different way to cope with complexity. A way that amplified transparency and put decision-making at the point of contact with the problems, and into the hands of Agile teams.

Chapter 3 – Benefits of Agile

Using Agile practices has extensive benefits in addition to those discussed in the last two chapters, especially in the safety-critical medical device field.

Culture of Learning

Agile practices are iterative rather than linear. The ability to develop iteratively fosters a learning culture in all disciplines (mechanical, electrical, software). Each new junction of the disciplines provides a baseline visible to all groups, which fosters peer-to-peer collaboration instead of hierarchal management structures.

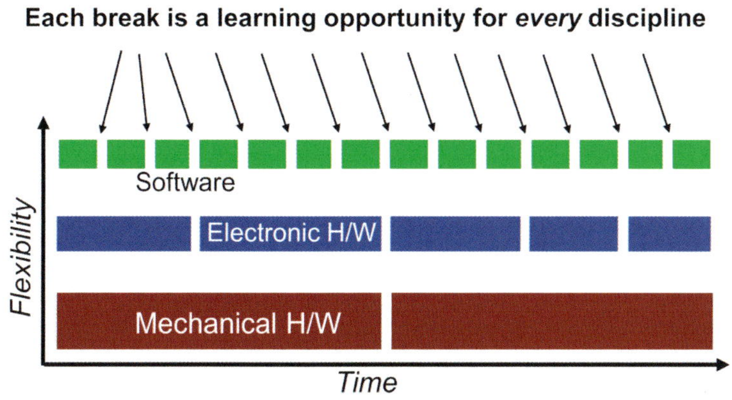

Each break is a learning opportunity for *every* discipline

Figure 5 Differing cadences for different disciplines in development

In Figure 5, the upper timeline represents elements like application software, system software, and user settings; all those things are easily changed, as they are pure software. The next timeline shows electronics, which might include reprogrammable chips, user-reconfigurable physical items (switch settings, socketed modules, etc.), and, of course, the circuit boards. The third timeline represents another class of hardware, which can include special components, sensors, housings, and designed materials. Why break hardware into two classes? You may not need to. This diagram illustrates a case where the mechanical hardware is more difficult to change than the electronics.

Even if the hardware iterations are longer than the software iterations, the goal is still the same: to get early feedback on possible solutions. As with software, the key is to have working hardware at each iteration boundary. (Note—Agile can be practiced without iteration boundaries, in a continuous flow. One such method is Kanban, which we'll touch on very briefly in Chapter 5.) That means each time the software has a new increment tested and ready, there needs to be workable hardware to integrate it with. The product has to be tested as a whole at regular intervals, because tested, working product increments are the chief measure of progress for Agile companies. That doesn't mean a new public release every few weeks; most organizations use these increments for clarity on their internal progress, and they save up several of them for a public release at extended intervals. For a medical device product, of course, a public release requires the extra work of a regulatory submission, so the internal "releases" will be much more frequent than the launch to market.

All Disciplines Learn at a Faster Pace Even When Only Software is Agile

In Figure 5, you can see that the software goes through three increments (also called sprints) during the time the electronics stays constant. Clearly the software team has more chances to discover a problem or a new idea—and act on it—than the hardware team. Therefore, during those three sprints, it seems the hardware team is not getting any benefit from Agile. However, experience has shown differently.

(Nancy) Let's use an example from the spectrometer project that my early Agile team built. Our hardware people were not attempting to be an Agile team. They were using long iteration cycles typical of traditional engineering planning. The embedded software team was Agile. We noticed that the math algorithm was sometimes failing to complete even when the raw sensor data was good—something that shouldn't happen. Because the software had Agile unit tests and system level tests that could run on target hardware and stable PC hardware, we could within minutes be certain that our software code was a perfect match to our mathematicians' model (which ran to completion when fed the identical raw data that we could capture from the sensor in real time). The electronics engineers looked at the problem and found a voltage point on the circuit board was far from the expected range. Substituting a different board fixed the problem.

Software could immediately add a range check on that voltage test point in the next sprint. But what range was within spec? The electrical engineers weren't sure. Since everything was new—the sensor, calibration, circuitry, even the science itself—they hadn't established clear bounds. So in the first sprint (for that period in the project) we added monitoring to show the typical behavior of the voltage, and in the next sprint we added trial warning bands and a message to the logging record. Finally, in the third sprint we had collected data from several instruments and situations. All the while, the hardware hadn't changed.

Next came a hardware upgrade that could incorporate much more understanding than it could have done if the software sprints were the same length as the hardware sprints. This is how the learning opportunities in any one discipline can potentially benefit all the engineering disciplines involved.

Meeting Legal and Regulatory Obligations

Another critical benefit of Agile is the ability, even necessity, to document incrementally. If you design medical devices or write software for clinical trial data management, you're working in the regulated world, with a number of legal and regulatory obligations around safety risk and documenting requirements and design, as well as providing traceability from risk mitigations to requirements, from requirements to tests, and from requirements to design.

A multi-level framework surrounds every development project you carry out. Standards set forth overall expectations for the industry: ISO 13485 and 21 CFR Part 820 set out requirements for your quality processes, ISO 14971[21] maps out steps for evaluating and minimizing safety issues, and IEC 62304[22] describes lifecycle elements for quality in medical device software (as well as the requirement for traceability of risk mitigations).[e] Standard Operating Procedures (SOPs) within your company lay out your internal framework for accomplishing those things in the context of your product area, and the projects you carry out generate the required outputs.

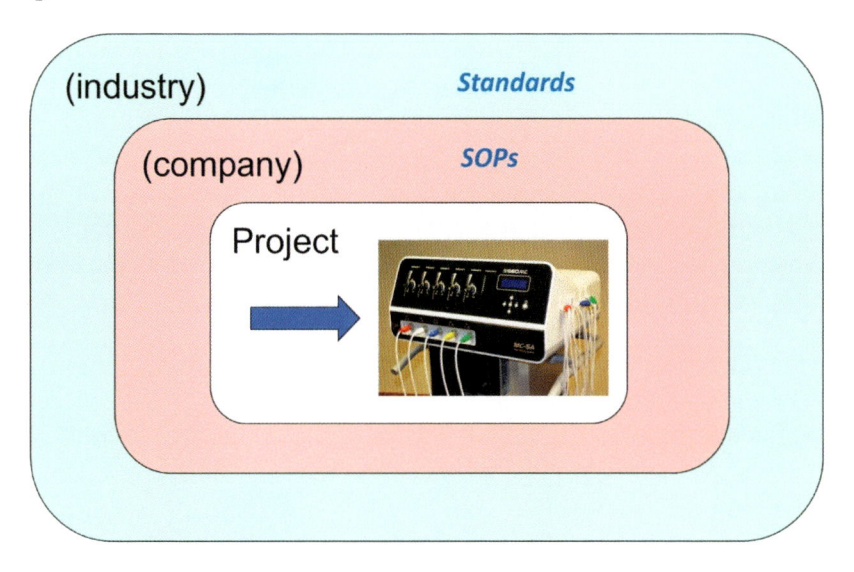

Figure 6 The framework in regulated product development

It's sometimes said that you actually create two outputs: products and documents about products. Neither is complete without the other. Those documents must serve the needs of several audiences: reviewers, users, and support teams (who often have to address product issues years later). Just don't let those obligations weigh you down; development tools (story managers, bug trackers, and automated test systems) can be used to keep track of all this information. Apply discipline to keep your information up to date. The key is to capture this information in the normal course of work you're already doing, rather than leaving the documentation until later, when it's always more difficult and burdensome to generate. We'll discuss this further in Chapter 7.

---

[e] FDA has also published a number of guidance documents on software on its web site – these can be searched at **https://www.fda.gov/MedicalDevices/default.htm**

Agile at its core focuses on flexibility and adaptability. Those words are often enough to make those in quality assurance and regulatory affairs squirm in their seats. How can a team be flexible and adaptable and still generate the right documentation? After all, for medical devices, "not documented" means "not done."

Your documentation is your proof that you said what you would build and then built what you said you would build. To satisfy regulatory requirements, you need evidence that you accomplished certain tasks and that you followed basic quality practices. You need these documents to be complete, consistent, and unambiguous—with hazards evaluated and mitigations defined, with traceability established, and with information sufficient for maintenance (which may be needed years later). Agile practices and ongoing discipline help fulfill these obligations.

What About Risk Management?[f]

Risk management is a subject of its own. We can only treat it briefly here, but Agile makes this easier, too. The features we aim to develop start out high-level and approximate, and as development proceeds, these general features are refined in order to implement them. So too does our general understanding of hazards become more specific and refined, allowing us to develop specific risk mitigations as a project moves forward. ISO 14971 §3.1 states that the manufacturer "shall establish, document and maintain throughout the life-cycle an ongoing process" for analyzing, evaluating, and controlling risks. This ongoing process fits neatly into the iterative learning environment that Agile provides.

---

[f] Here we mean safety risk management, which is the focus of ISO 14971. Project risk is important, of course, but not the subject of this discussion.

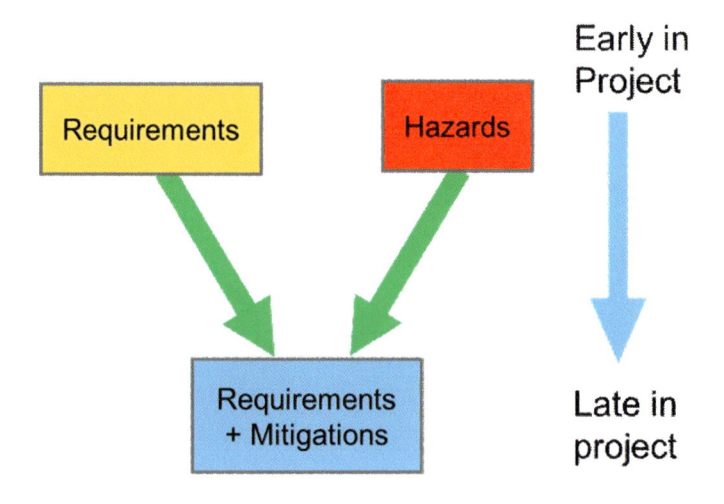

Figure 7 Risks vs. Requirements – Ongoing Refinement

Built-In Quality

Agile offers benefits beyond a smooth way to document during development. It also offers quality, and high quality is critical to medical products. At a conference some years ago, I (Nancy) met a technical lead who had moved his firmware team to Agile practices for their company's medical products, and I was eager to know how much improvement he saw in the number of defects that got out to customers. "None," he said. Then he explained that I was asking the wrong question: He'd seen an improvement in his time to project completion. Everything went through their final verification stage and it always came out clean. To be clear – it stayed in the final verification until it was clean, however long that took!

The difference Agile made was that the verification stage no longer wrecked their schedule, because the product entered that stage with far fewer defects than in the past. In fact, every stage of their process now sent fewer defects to the next stage. In their older process, the defects accumulated after each stage of the work, so there were more defects entering verification that had to be cleaned out, and troubleshooting is not predictable. Naturally it was less work to clean a smaller number of defects. Figure 8 shows this effect.

As evidenced by this example, testing at frequent intervals (typically every two to three weeks) creates a huge advantage. In addition to reducing the grand total of defects, the rework cost of those extra defects falls to zero. The cost per defect is dramatically higher when more defects are present, because when you have two defects interacting, it's much more than double the work of solving only one.

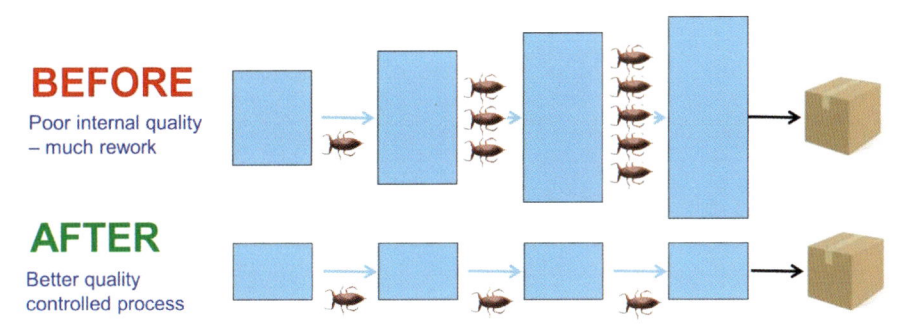

BEFORE
Poor internal quality – much rework

AFTER
Better quality controlled process

Figure 8 How Agile improves throughput for a high quality product

Agile software technical practices[g] do a superior job of keeping defects out of the codebase, and quickly finding those that do get in.

Table I. Measured Defect Densities for Agile vs. Traditional Projects

Team	Defects / FP	Process
Follett Software[23]	0.0128	Agile, XP co-located
BMC Software[23]	0.048	Agile, Scrum distributed
GMS[24]	0.22	Agile, XP for embedded
Industry Best[25]	2.0	traditional
Industry Average[25]	4.5	traditional

The Software teams whose defect performance is tabulated in Table I all had collected sufficient data for their defect rates to be directly compared to each other. The traditional teams were studied by

---

[g] Largely the practices advocated in Extreme Programming, or XP.

Capers Jones, with "Industry best" being those at the 95th percentile for defect-free code. Nancy's early Agile team building the spectrometer project achieved 0.22 defects per Function Point, and the figures for BMC and Follett were analyzed and published approximately seven years later by Michael Mah.[23] The spectrometer team performed nine times better than the best traditional teams, and Follett Software performed seventeen times better than the spectrometer team! The Follett project was a PC-based application for libraries, not real-time embedded new development, but it's still a very impressive leap.

Technical Debt: Why Does Software Become Difficult to Change?

As features are added to software, defects are added too. When you don't test the code at frequent intervals, many of these defects remain hidden for a long time.

Other things happen, too. Methods become longer (and harder to understand), variables start to take on meanings that differ from their original intention, and classes designed for a single purpose are used for additional related purposes. There is an accumulation of small cleanups that need doing, and at some point they create inertia in the codebase that makes it much more likely that any new changes will also have a much higher defect rate.

We call this accumulation of small-jobs-not-done "technical debt," because you can't put off the cleanup forever. You are going to have to do it eventually, and while you postpone that day, you are accumulating interest in the form of the extra defects you create and the fact that the codebase is simply harder to work with.

We tend toward an assumption that to do twice as much of something requires twice the amount of effort. That kind of linear assumption is trouble! Software has a very strong non-linear nature: Adding a simple feature when the codebase is small might take you a few hours, but when the codebase has grown to ten times that size, adding the very same feature could take over six times the effort! This effect is much stronger in software than in other types of engineering work, and it's a big reason that estimating the time necessary to create new software (which we'll talk about later) is so difficult.

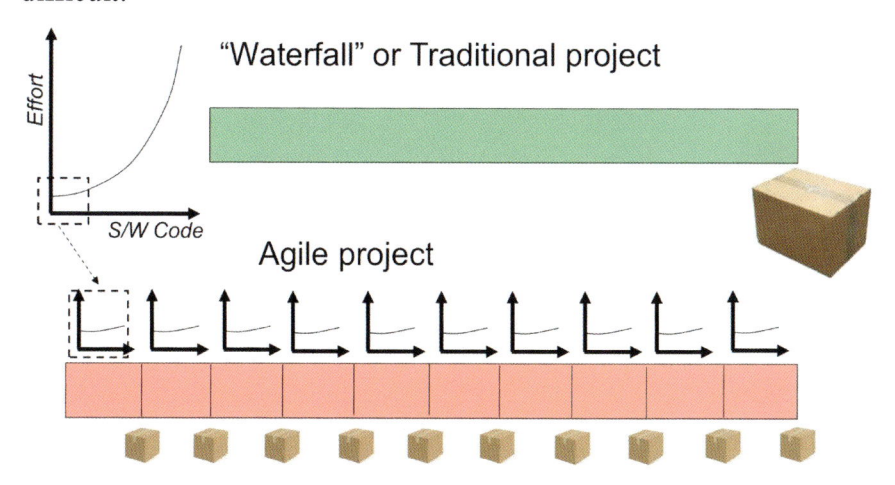

Figure 9 Regular, thorough testing in Agile projects minimizes effort cost per unit of completed software

In Figure 9, the upward curving graph of effort vs. software code expresses this effect. Essentially, all the features added later on to a traditional codebase cost significantly more money to build, and that depends merely on how big the codebase has grown by the time the feature is added. It's not the size of the codebase that is the problem, but the amount of technical debt it contains. If the codebase is clean, it can be large and still be easy to work in.

Agile teams stay on the low-cost part of the effort curve by thoroughly testing the code at each sprint and by keeping technical debt low. This effect is illustrated in the lower half of Figure 9.

Flexibility – Responding to Change

Another of Agile's advantages emerges in its ability to adjust and respond to change mid-stream. Most companies use a "phase-gate" process to review and control their development projects. The typical phase gate process mandates reviews at a series of stages within development: requirements complete, architecture complete, alpha testing, and final implementation and verification. Specific rules govern the ability to pass through each of the gates. Because these gates require you to have sections of work done in big "chunks," they usually make it difficult to work in a flexible, adaptable, Agile way. If new feature demands arise in the middle of a development project, they can cause chaos in a traditional system. Requirements need to be rewritten, re-prioritized, and approved. Architecture and detailed design might need to be revisited. And so on. With an Agile approach, important new features can simply be prioritized in the next sprint planning meeting.[h]

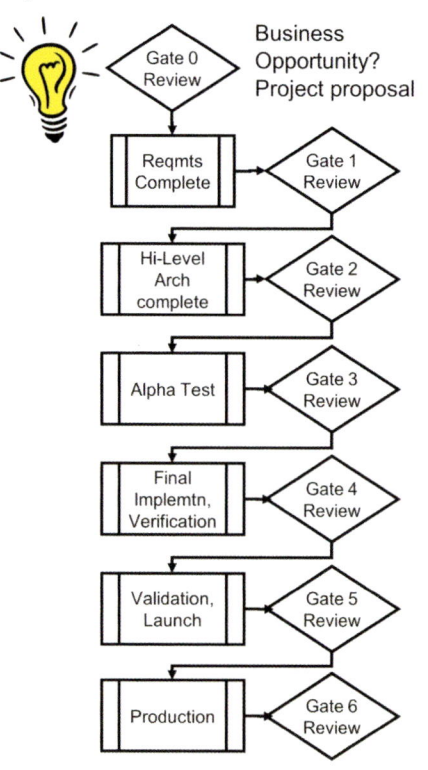

At least initially, the Agile approach represents a fundamental departure from standard product development, and you may need to include it in your development plan.[i]

Figure 10 An example phase-gate process

Refactoring

Code refactoring is the process of restructuring existing computer code—changing the internal workings of the software—without changing its external behavior. Refactoring improves nonfunctional attributes of the software. Advantages include improved code readability and reduced complexity; these can improve source-code maintainability and create a more expressive internal architecture or

---

[h] Naturally, new and changed requirements must be approved, to comply with design control.

[i] IEC 62304 appendix B (section B.5.1) discusses software development planning.

object model to improve extensibility. Refactoring is another practice that Agile software teams use to keep the code free of needless complexity and confusing aspects that tend to grow over time. Refactoring is a way to preserve the present behavior of code while improving its internal structure. In other words, it pays down the technical debt and makes features less costly to add. The details are outside the scope of this book, but we recommend *Refactoring* by Martin Fowler[26] for further information.[j]

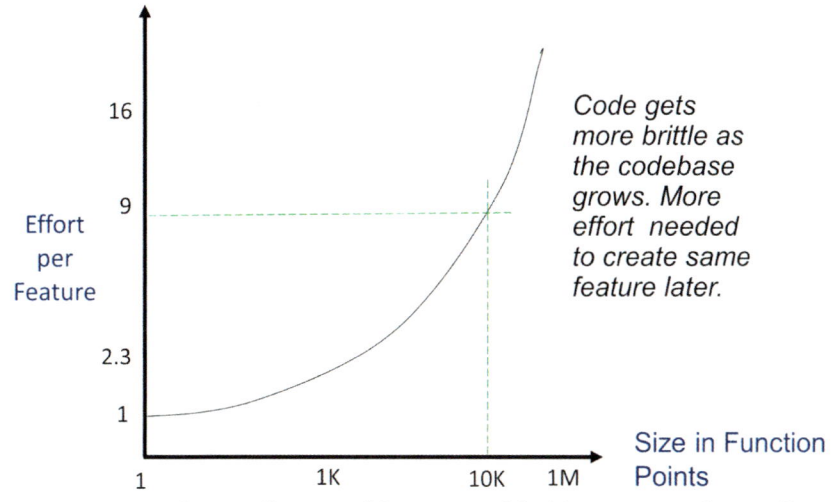

Figure 11 Cost inflation of features added later to traditionally developed software. (Data from Capers Jones[27])

Research data has quantified how the effort for the same work increases with the growth of a codebase, as shown in Figure 11. Bear in mind that this is data from code where the teams were not using Agile practices, so it is showing the additional cost that results from both defects and technical debt accumulating in the codebase.

By cleaning the code at frequent intervals, you work this effect in the opposite direction: You build your software for a fraction of the cost that non-Agile companies are paying to work in a codebase of the same size.

---

[j] Once change control has started, a formal change request must be initiated for changes to code. A firm may, however, determine how granular formal change requests must be, to minimize how onerous this process will be.

One mistake often made by self-taught Agile teams is that they don't test thoroughly within each sprint. Without understanding the necessity of regular testing, and without support for this crucial step from their management, they are too pressed for time to refactor adequately. It does take time to become good at refactoring, at test-driven development, and in using Agile regression testing. But those practices are key for keeping the codebase in good condition.

Decreasing Time to Market

Agile also offers an additional financially valuable benefit— it decreases time to market. This is done in two ways; developing minimum shippable features (or minimum marketable features— MMF) and avoiding unnecessary work.

Under the Agile approach, each time your team adds a set of features, the product needs to be runnable. In classic Agile terms, this is called "potentially shippable." Indeed, in one medical device project reported at the Agile 2009 conference, management decided to move ahead with regulatory filing even though a number of features had not yet been implemented. Their comment was "At time of commercial launch, a number of features, once thought to be essential, were not included. Some were deferred as long as three years. Nonetheless, the product was considered highly successful, and trading off nice-to-have features for three years of sales is an easy choice." They realized that the product was complete enough to serve the end users ... and the team continued developing a more complete version to release later. Flexibility like this simply isn't available in a traditional development model.[28]

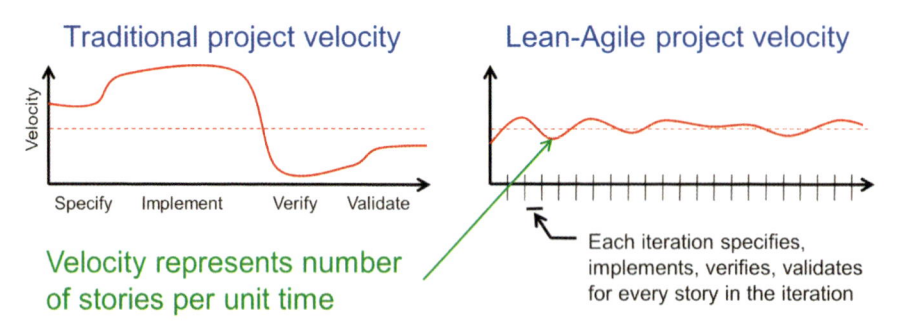

Figure 12 Evening out velocity makes predictions and estimates easier

However, the biggest leverage of Agile is the amount of work *avoided*. Agile teams don't work faster than traditional teams. If anything, they may appear to be working more slowly to a casual observer. The leverage comes from identifying waste early and eliminating it. The defects discussed above are pure waste. Catch them early! So are features no one will use, and those that customers can't figure out how to use. They should be dropped before effort is put toward them.

Traditionally, software teams could claim rapid progress early as they built separate pieces of their product. But all that speed would come crashing down when it came time to integrate their work (shown by the steep downward slope in the left side of Figure 12). Agile teams insist on integrating continuously, which allows them to work at a nearly constant pace.

The steady pace alone offers many benefits. First, it lessens the need to add or drop people, saving you money. Second, it keeps the stakeholders from panicking! Third, it makes estimating easier by creating a closer correspondence between elapsed time and quantity of work (which may be simply the number of stories or a sum of story points). It also offers a baseline that makes the effect of an experiment easier to see.

The Agile Advantage in Cybersecurity

Adaptability, flexibility, clean code—while these attributes all lead to financial benefits, they are absolutely critical in another area. In 2017, investigators at San Mateo-based TrapX Security detected malware on medical devices at major healthcare providers across the globe. They found malware planted on several types of medical devices including an x-ray printer, an oncology unit's MRI scanner, a surgical center's blood gas analyzer, and a health care provider's PACS (picture archiving and communication system).[29]

Similar incidents elsewhere have alerted the FDA that cybersecurity is a serious issue for medical devices.

The agency has held several public workshops and webinars on cybersecurity concerns and has issued guidance documents (a) requiring medical device manufacturers to develop cybersecurity controls in the design phase of their product development, and (b) recommending post-market actions to deal proactively with cybersecurity vulnerabilities. New device submissions are not even accepted unless the manufacturer has completed a cybersecurity evaluation.

The urgency of the cybersecurity situation appears to be running headlong into industry complacency, according to remarks by Dr. Jack Lewin, Chairman of the National Coalition on Health Care at the 2018 Software Design for Medical Devices conference. He stated that in confidential conversations with industry CEOs about cybersecurity, he was told "We'll have to wait until we are forced to do it" and "We will have to wait until it becomes an economic imperative."[30]

Our modern medical device industry has the same attitude toward cybersecurity that the automobile industry of the 1940s had towards vehicle safety! Cybersecurity is an evolving issue, yet a number of reports suggest that insufficient effort is being undertaken by the medical device industry to address the risk of attack.[31]

We contend that Agile teams can achieve software defect rates far lower than traditional teams. Code that is more defect-free is certainly better for security, but we believe that Agile teams can be used more strategically to provide further protection. The core Agile practice of using cross-functional teams is a vehicle for helping every function to understand how their area can build security in; we cannot afford to have security be just another knowledge silo. All too often, securing a medical system has meant creating a stronger moat while still leaving all the castle's interior doors unlocked. Security cannot be as good when added afterward, from the outside in.

In the same way that Agile methods build quality in from the start, Agile teams can (and should) be used to design security at every level, from the inside out. This means at the hardware layer, the firmware, and the systems software, as well as at the application software level. Cybersecurity researcher Hannah Murfet has offered specific recommendations for each stage of the software development lifecycle:[32]

- Risk management is not a one-off activity and should consider security as well as safety.[k]
- With evolving threats like cybersecurity, there is benefit of a continuous risk management review throughout the full product lifecycle.
- Management support is essential to drive best practice and release controls. Security is as important as new features!
- Training and conferences are important mechanisms to maintain competence and share industry best practices.

When our medical device industry takes this thinking to heart, its cybersecurity will start to come to the level necessary for the 21st century.

Findings from Actual Business Experience

(Brian) Even with all the benefits we've described, I've often encountered deep skepticism about Agile among quality assurance and regulatory affairs managers. One QA/RA head actually remarked to me, "I know that TIR 45 exists, but I'm still not convinced about this Agile stuff." This was quite a statement, considering that AAMI Technical Information Report 45[33] outlines how an effective Agile approach is consistent with the safety and quality goals in medical device development. Nevertheless, it turns out that Agile *is* being used successfully on regulated medical projects. For example, Given Imaging (now part of Medtronic) has for a number of years employed Agile in development of its "pill camera." GE Healthcare not only pursues Agile development for its imaging system work, but has also set up an entire demonstration center where the company brings in physicians to participate in periodic demonstrations. Device makers Dräger Medical, Elekta, and Renishaw have also adopted Agile approaches for development. Medidata Solutions, a mainstay in clinical trial data management software, follows an Agile approach for its clinical trial software, and outsourced development firms Systelab Software and Synchroness are using Agile methods to greatly benefit their device and diagnostic clients. These are only a few in a growing field.

---

[k] As well as data privacy, per HIPAA and the EU's GDPR regulation.

Despite these successes, Agile is not plug-and-play. Agile represents a significant change from the old style of command-and-control, top-down project management. The Learning Consortium, a group of eleven companies formed to share information on making every business function Agile, reports that Agile is a mindset more than a process, a technology, or a structure. They suggest that company-wide Agile transformation can take five or more years, and that it requires strong, inspirational leadership with leaders that have a deep confidence in Agile principles.[34]

Harvard Business Review has reported that Agile methodologies "are a radical alternative to command-and-control-style management,"[35] and radical change is hard. However, we firmly believe that adopting an Agile mindset and development process is worth this effort. In the next chapter, we take a closer look at the specific practices and mechanisms of Agile.

Chapter 4 – Applying Agile: The Importance of Stories

While there is no one right way to implement the Agile principles, this chapter describes some common Agile practices that have proven to be a good starting point for work that involves hardware and software in a regulated industry. Some Agile teams work using iterations (often called sprints), while others work in a continuous flow. Some have one person (Product owner) in charge of deciding what features to implement, and others have a team making those decisions. Sometimes such variations make a critical difference. While Agile is very adaptable, it's also very situational; in addition, experience matters. That is why we recommend coaching. This chapter describes a safe starting point for medical device work, and leaves other approaches for a different book. This is a primer, after all.

TENS – Our Example Project

Throughout the remainder of this book, we'll use an example project to illustrate Agile approaches: the development of a TENS device (transcutaneous electrical nerve stimulation) for clinical (not home) use. This example will help illustrate various crucial elements of Agile.

TENS is a pain-relief therapy in which weak oscillating electrical signals are applied to a patient via standard skin electrodes. The goal is for treatment to be fully automated: working parameters are to be set dynamically, with no manual adjustment required other than regulating stimulus intensity, which is manually set at the perception threshold. The first step in developing this device is breaking the various features into "stories."

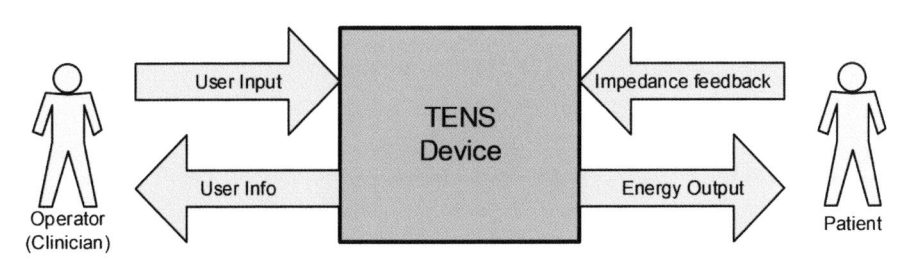

Figure 13 TENS Technology

Stories are the "Kernels"

(Nancy) Agile stories are nuggets that capture what is to be implemented, why it is important and to whom, and what actions will demonstrate that the feature is working correctly. Well-crafted Agile stories provide a beacon lighting the way from idea to action.

(Nancy) I've seen teams that clipped their old-style "The system shall…" requirements document into paragraphs and called each an Agile Story. This won't work. Agile Stories are not specifications. They are a record of a conversation that can still continue as needed. Below is an example of taking a laundry list of features and turning one of them into a story.

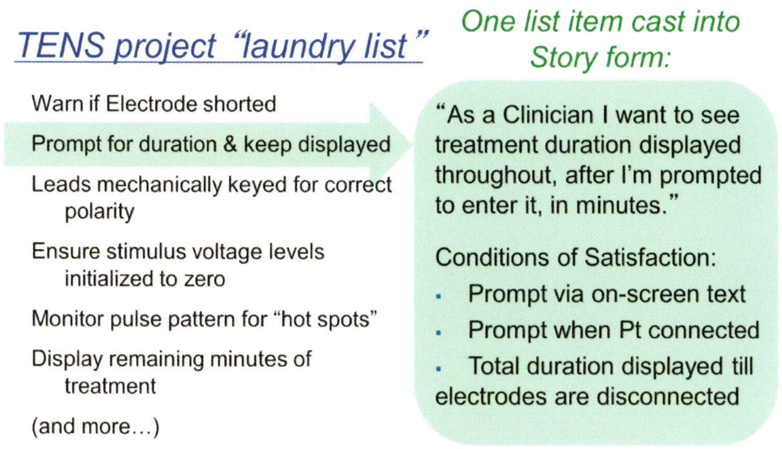

TENS project "laundry list"

- Warn if Electrode shorted
- Prompt for duration & keep displayed
- Leads mechanically keyed for correct polarity
- Ensure stimulus voltage levels initialized to zero
- Monitor pulse pattern for "hot spots"
- Display remaining minutes of treatment
- (and more…)

One list item cast into Story form:

"As a Clinician I want to see treatment duration displayed throughout, after I'm prompted to enter it, in minutes."

Conditions of Satisfaction:
- Prompt via on-screen text
- Prompt when Pt connected
- Total duration displayed till electrodes are disconnected

Figure 14 Sample story from the TENS backlog

The version of this story on the right-hand side above is more *useful* in helping the team to actually achieve what they set out to do. This version is an end result of a story crafting session. Story crafting isn't a set of rules, or even a defined set of questions. It's like a conversation, but with some goals along the way that need to be visited. In fact, the Agile story itself is not as important as the steps you take to create it. A story like the sample one shown for the TENS product comes out of a conversation between people with very different perspectives on the feature described. In this case, it would ideally include a clinician, a product designer, a software developer, and a tester. Together they understand the whole story and can agree on what will demonstrate that the system is behaving correctly after the story has been implemented.

Properly worked out, an Agile story captures a number of different elements: what to build, an estimate of the effort needed, any plans for risk mitigation, the approach to testing, and elements that QA can use to ensure documentation and traceability.

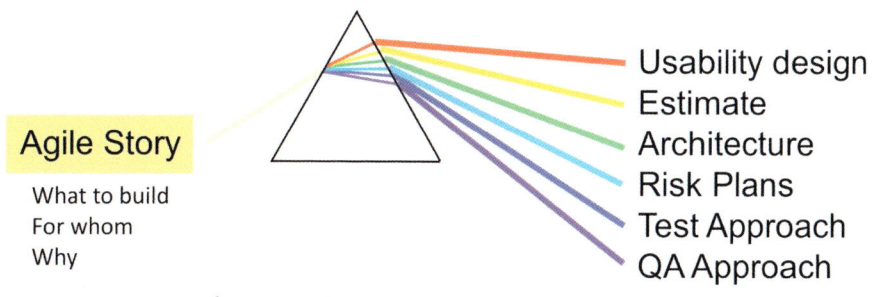

Figure 15 Elements that emerge as an Agile story is brought to readiness

The artifacts shown at the right of Figure 15 are not part of the Agile story—the story provides a foundation for incremental growth of each of them. Their details have a home in other documents that may be linked to Agile stories.

These stories provide the basis for all sorts of Agile practices, from setting iterations to delivering earlier and on a regular basis. Each story provides a vision for one strand of the work to be done. These stories can also be used to estimate the amount of time or work it may take to complete a project and to estimate progress (more on measuring progress later).

Estimating Work Based on Agile Stories

The point of any estimating is to say with some confidence how many days from now the work will be completed. Of course, clarity on what "the work" entails matters—that's the job we do when writing user stories. Whether we focus first on some other aspect of the work, such as its complexity, number of interfaces, or expected lines of code when completed, the object is to end up with a quantity of calendar time till delivery of the work.

(Nancy) When coaching teams new to Agile practices, I often find that using story points provides a fast transition to Agile estimating, because the numeric basis is familiar. The abstract nature of the units can be picked up quickly when combined with that familiarity.

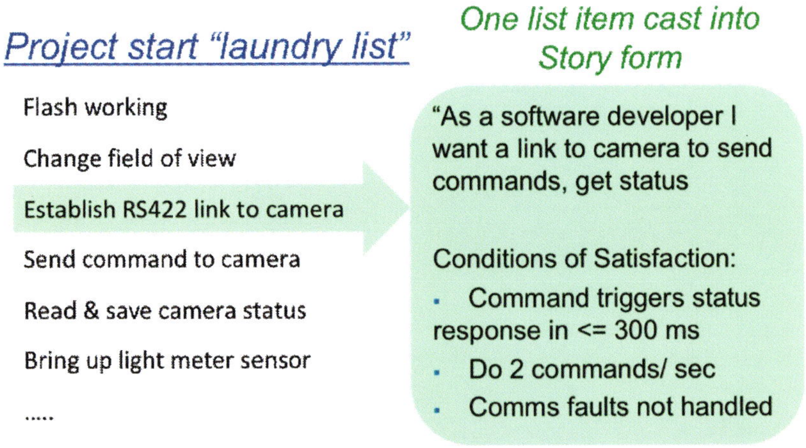

Project start "laundry list"

- Flash working
- Change field of view
- **Establish RS422 link to camera**
- Send command to camera
- Read & save camera status
- Bring up light meter sensor
- .....

One list item cast into Story form

"As a software developer I want a link to camera to send commands, get status

Conditions of Satisfaction:
- Command triggers status response in <= 300 ms
- Do 2 commands/ sec
- Comms faults not handled

Figure 16 Developing an item from the backlog into story form

But why not just estimate in days? It's a frequent question. The problem is that even if you can estimate your work well, there are many interruptions that are hard to control or factor out of your estimate. It's better to break the problem down a bit. Let's walk through an example of estimating stories that might be at the start of a medical device project.

These stories could be a portion of several different kinds of medical device; it doesn't matter for the purpose of describing how to estimate them. Once a story has been fleshed out through conversations between those who know different aspects of it, we'll have something that looks like the "Establish RS422 link to camera" story shown in Figure 16. The conditions of satisfaction give important scoping information: Apparently there was already some discussion about whether to include communications fault handling logic (which would mean the story is much more work), and they decided to leave that till a later iteration.

Story Point Estimating

So, how many story points are in the Figure 16 story? Story points don't have any inherent meaning. Some people define them to mean a person day, or half a person day. But that is simply giving another name to a unit that already exists. Suppose instead that we let a story point mean a unit of inherent complexity that is in the story. This isn't really such a strange thing to understand.

Consider this example from a history book about the 1660's:

In Albany, New York, one beaver pelt could buy six jugs of brandy.

In Montreal, Canada, one beaver pelt could buy one jug of brandy.

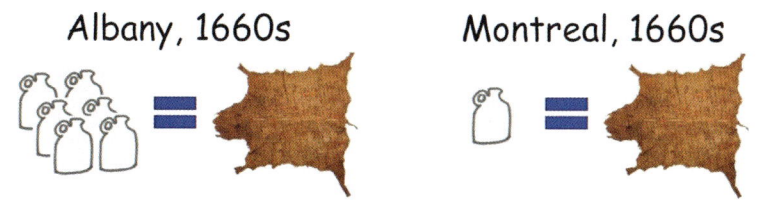

Figure 17 Exchange rates for pelts and jugs of brandy

The point is that there is nothing inherent in the pelt or the brandy that makes this equivalence work. It is simply a social convention. We can make a social convention for measuring the size of software. That is what "story points" are.

It helps to look at more than one story so that comparisons can become clear. Before trying to estimate the story points on the very first story in your project, it's good to flesh out several stories to the same level as shown for the RS422 story. Before doing any estimates, have the team write these stories on cards and arrange them on a table with the smallest at one end and largest at the other.

Smallest → Biggest

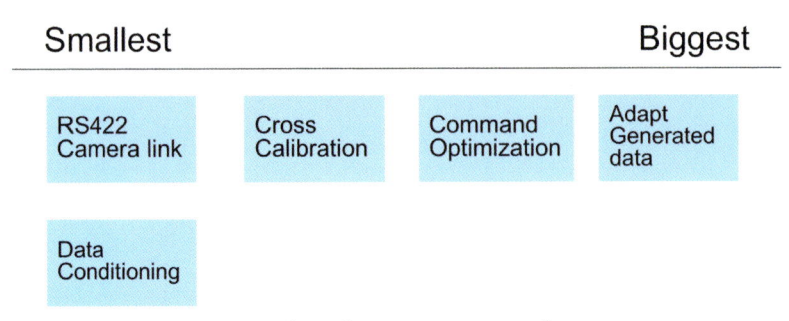

- RS422 Camera link
- Cross Calibration
- Command Optimization
- Adapt Generated data
- Data Conditioning

Figure 18 Rank ordering a group of user stories

At this point it's good to have a variety of sizes represented. If the stories are clustered together in size, then develop some more of them (clarify their purpose and conditions of satisfaction) so that there are several different sizes of story. In the figure, the team decided that two of the stories are of equal size, and both are small. So those are arranged vertically. However, this does not give us an indication of how much bigger the next size story is compared to the smallest ones.

In order to quantify how much bigger the larger stories are, Planning Poker is a helpful exercise. Here is what the outcome of that may look like:

Story points: 1, 2, 3, 5, 8, 13

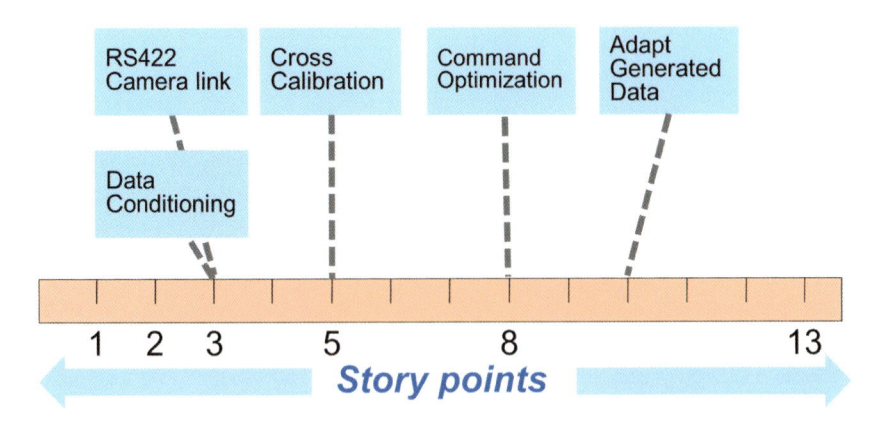

- RS422 Camera link → 2
- Data Conditioning → 3
- Cross Calibration → 5
- Command Optimization → 8
- Adapt Generated Data → 13

Figure 19 Quantitatively ranking user stories via Planning Poker

The following is a good way to start if the team has not done story estimation at all, or if this is a new project where no stories have been estimated yet.

Choose one of the small stories and give it a small number of story points, without doing any estimate at all. It's a seed for sizing all the other stories. In the example, the team chose "3" as the value for the RS422 story.

"Planning Poker" is an iterative approach to estimating, and a participatory decision-making process for the whole team. Majority vote is the default decision rule.

Steps in Planning Poker

- Each estimator is given a deck of cards, each card with a number written on it
- Customer/Product owner reads a story, and it's discussed briefly
- Each estimator selects a card with an estimate
- Cards are turned over so all can see them at the same time (independent estimates)
- Discuss differences (especially outliers) – determines only the info that matters
- Re-estimate until estimates converge (rarely need more than three rounds of this)

The next step is to choose a slightly bigger story—Cross Calibration—and ask the team this question: If the RS422 story is three points in size, what is the right size for the Cross Calibration story? Their cards will have numbers that have gaps, e.g. 1, 2, 3, 5, 8, 13, to help them avoid dithering over numbers that are only different by a small percent.

By discussing each story with the set of people who will implement it, the information necessary to understand why it's a big job (for instance, it might be extra difficult to test), or why it's a small job (someone knows of code from another project that can be re-used for part of the work) is drawn out from the people who know the most about it. The beauty of this method is that's the *only* info that comes out. Nothing more than what they need for the estimating.

The main mistake people make with Planning Poker is spending too much time discussing things. It takes some discipline to limit discussion and move to a re-vote. A team that becomes good at this method can move along quite well, taking less than ten minutes per story on average.

James Grenning, the inventor of Planning Poker, has moved to using a different method now. He likes "affinity grouping."[36] That is a method where the team will take turns placing the stories in groups, without any discussion initially. At a later point they add a numbered range as shown in Figure 19. As they continue a few rounds of turns, there may be a card that one person moves right and another moves left, so that card keeps jumping back and forth. The team will have a short discussion about why there are different views about that story's size. Note that on each person's turn, they can make only one move—either turn up a new story card and place it, or move a card on the table to another spot.

If a story is a thirteen or bigger, then the team takes that as a signal to break the story down into two or more smaller stories. There is a reason to keep stories within a one-to-ten ratio of sizes. It's based on lean flow concepts, which we'll discuss briefly in Chapter 10.

T-Shirt Sizes Estimating

Some people use a simpler range of values for their story sizing, although they're still aiming for an overall sense of the story's difficulty. Instead of numbers, they use T-shirt sizes of S, M, and L, and they may add XS, XL, and so forth. They may use discussions like those in Planning Poker or the affinity grouping technique to quickly get the team's opinion about the size of the story.

Getting a Time Estimate from Story Points or T-Shirt Sizes

There is one more step to do if you're using a new unit like story points or T-shirt sizes for estimating; you still need to arrive at an estimate in days or hours so that you can tell stakeholders how many stories will be completed in the iteration. One way is to just work for one sprint without making a promise of what will be completed.

Based on the number of story points completed, you now can expect that is approximately the team's capacity for one iteration.

Another way is to use traditional estimating to predict how many stories from the top of the backlog (most desirable stories) can be completed given the number of labor hours the team has for the iteration.

Calendar Days Estimating

Estimating calendar days can still be done in the right environment. Menlo Innovations is a company founded on Agile principles and practices.[37] They use pair programming for all of their work, and everyone is co-located. They do their estimating in days and hours. It's very simple! There is no prohibition against this — any Agile team could do likewise. The issue is that in most companies moving from traditional development, people still have many interruptions to deal with, and that's what poses a challenge. At Menlo, they can work all day with practically no interruptions. That makes it much easier to estimate directly in calendar time.

No Estimates

Some companies don't use estimation at all. They apply all the labor to working on the stories. If there is a need to predict when a particular story will be done, they can observe the rate of flow and make a quick projection, just as you might do to give an arrival time if you're driving in heavy traffic.

Whether to use estimates at all can hinge mostly on how engineering interacts with upper management, rather than how well estimating helps the engineering team. If management needs to report to external clients, or if management is not yet comfortable with the Agile approach, providing estimates and meeting them helps build the necessary trust for a move to Agile to be successful.

Chapter 5 – Applying Agile: Iterative and Incremental, not Linear

Deliver Early and Often

If you've done any amount of product development, you've experienced the traditional method: marketing writes up a specification, management assigns it to development (on a timeline that's always too short), and the whole mess is "thrown over the wall" at engineering, which struggles to build the entire thing and deliver it. Agile replaces this broken model with one of regular, frequent deliveries.[1]

One of the crucial benefits of Agile is that its principles permit these small deliveries: something runnable, early and often. "Runnable" may not mean a complete product, but it must have enough architecture to operate and be used for demonstration. Let's call these increments vertical slices, since they involve something at every layer of a design (hardware, operating software, command system, user interface).

In traditional product development work, "progress to date" tends to be rather theoretical during much of the development process. It has to be computed based on what percent of the final product is completed, but if the project were stopped halfway through, it's likely that there would be nothing that could be delivered at all.

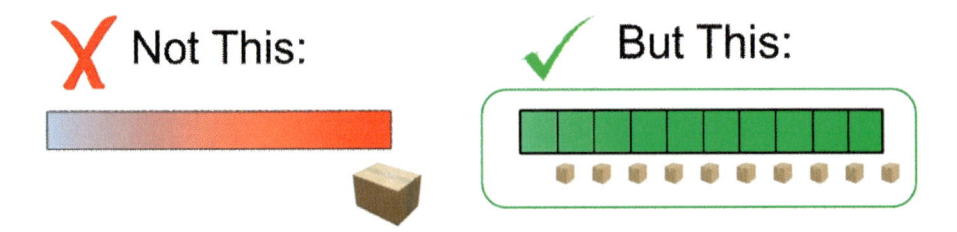

Figure 20 Small, frequent deliveries are essential in Agile

---

[1] The design control requirements, according to 21 CFR 820.30, still apply—we just believe they can be addressed in an incremental/iterative approach.

If instead it were possible to complete the development in stages such that each one had customer value and could actually be used, the progress to date would be transparent, and the project would start generating revenue earlier. If it turned out that customers didn't like the product, or if they wanted to change product features, we'd know earlier and with less at risk. This is the vision that Agile methods work to implement.

From an engineering standpoint, it often appears natural and efficient to build systems in horizontal layers: hardware, firmware, operating system, command system, user interface. The problem with this approach is that most features aren't ready to be used until close to the end of development. Instead, Agile's use of vertical slices allows complete features to be delivered within a runnable system and spanning all layers, which means you can receive feedback on them. As software and products have grown in complexity, this early feedback and transparency on progress to date are so valuable that they are well worth the effort of learning to work in vertical slices.

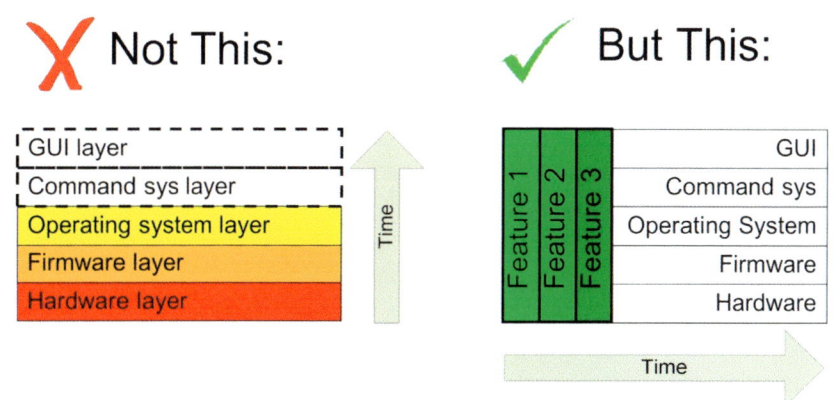

Figure 21 The concept of building in vertical slices

This vertical slice way of working can be done quite easily when only software is involved, but no one is going to build a multi-layer circuit board one layer at a time! Hardware gives us constraints on how far we can take the slices model, which we'll talk about later, but it's an important guiding vision.

Stories and vertical slices are closely related. One definition of user stories reads: "User stories are short, simple descriptions of a feature told from the perspective of the person who desires the new capability, usually a user or customer of the system."[38] A question that comes up often is: "When we define a story so it has value to the customer, it ends up being very large because of all the system layers it touches. But if we divide it down, then the smaller stories don't mean anything to customers." The problem when stories are too large is that you lose visibility into where the work stands. When stories have meaning to the customer, you can get more useful feedback.

If your first story in the TENS project says to bring the unit to "ready" state, indicated by an LED indicator being turned on, it is understandable by the customer, but it might be a three-week-long job for that one story because so many foundations have to be built! Here's a practical way to handle that.

Recognize that some customer-facing stories need to be broken down further into technical stories (also called building-block stories). So a story with the headline "Indicate that TENS is ready for treatment start" could contain these building blocks:

- Bring up data-acquisition circuit board
- Bring up main circuit board
- Port power-on self-test software from previous TENS model

This can be broken down even further. Rather than one story that says "bring up data-acquisition circuit board," which might take most of a week, consider a set of still smaller stories that don't slice all the way through the technology stack. The stories are shown as vertical blocks in Figure 22, indicating which layers of the technology they address.

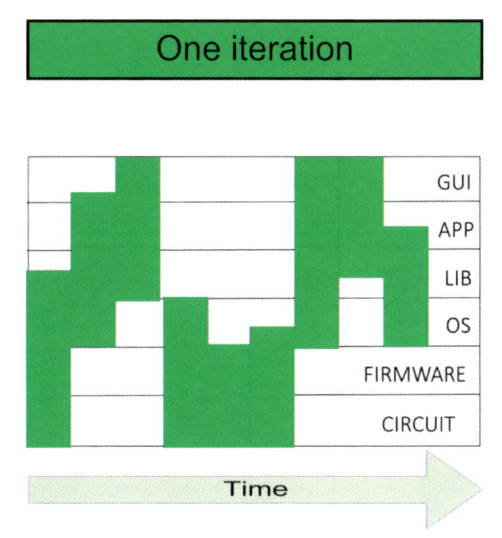

Figure 22 Not every story spans every layer in the system

The left-most story in Figure 22 can be about firing up the newly delivered board and installing the OS and some libraries. The next story might be checking out the rest of the OS, installing the rest of the libraries, and getting a "hello world" app to execute. The next story adds a graphics library and pipes the "hello world" text out to the mini-display on the board. And on it goes as you bring up flash memory, the floating point unit, and other hardware on the board.

This allows high transparency into the ongoing work. Embedded systems have deep technology stacks, and many of the tasks are simply never going to hold customer value as such. However, by using building-block stories we balance the need to keep focused on customer value against the need to measure visible technical progress. Agile stories also balance incremental feature build with a cohesive design vision. The book *User Story Mapping* by Jeff Patton[39] is a good source for more information on this, as well as greater depth on everything about Agile stories.

Control via Stories, not Phase Gates

As we mentioned earlier, we used to work to milestones (or phase gates) similar to those depicted in Figure 10:

Specification sign-off

Analysis complete
Code complete
Integration test complete
Project delivery complete

(Nancy) There are a few problems with these. A specification document that runs to hundreds and even thousands of pages (yes, I really did work with those before Agile came along) is nearly impossible to get signed off. Who can tell where there are gaps or contradicting statements that will certainly lead to battles later on? Searching out all the possible issues in a massive specification inevitably holds up work, causing delays at each phase gate.

In addition, "analysis complete" is meaningless. In the defense world, it generally meant "when the funds run out for the analysis phase." In the commercial world, it came to mean creating yet another big document. And the idea that we'd really finish all the coding before we tried to integration-test it is painful to even think about today. The reasoning was that our unit tests would catch most of the problems before integration test time. But the first time I ever saw genuine, full-up unit tests in use was on a project about five years before Agile that used some of the collaboration patterns of Agile. That project did it, and Agile teams did it. Every other traditional project I worked on had to skip unit tests because we were either out of time or out of funds.

Integration test was the first real milestone that could not be evaded. It was a real moment of truth, and unless you worked in a very well organized way, your project might not survive it.

Since the integration test is the only real test of whether you are going to deliver, why not have it earlier, when you have more options still available? And why not do it often to make sure you're still on track? Agile teams do that!

The Phase Gate type of milestones don't make sense for Agile projects. There are still what could be called natural states of development in projects that do make sense for Agile:

- Preparation - Training, ordering equipment, setting up test environments

- R&D - Getting key feasibility questions (tech, business) answered or you might not proceed
- Development - The period of cycling regularly through detailed design, code, test, and integration with regular customer involvement
- Final test - Satisfying a legal requirement or testing not feasible to do earlier, e.g., medical devices or an initial flight of an airliner
- Deploy - Releasing the product, possibly in a limited run to shake out the last issues

That would suggest we exercise control points at those stages, as shown here:

Figure 23 Phase Gates "lock" progress from one activity to the next

But that can create a new problem.

If we lock progress from one stage to the next to make sure that everything in the preceding stage is complete, we actually block the flow of value. This frequently happens in traditional phase gate systems. In an Agile approach, we unlock those phases and allow work to move forward at varying rates, and we control and monitor that work at the story level.

Figure 24 Agile removes the phase locks

Each story must proceed from preparation all the way to final testing before we can declare it done. Early stories may focus more on preparation and R&D content, but the point is still the same: we monitor the completion of stories to know where we are in the project.

The story level is the more granular point of control. Unlike a 300-page specification, stakeholders/customers can look at a story idea, engage in the conversation that develops it further, and make sure that they understand how it will behave—all because they can help define its acceptance criteria. And if there is a story they want to postpone, the rest are not held up. For example, our stakeholders might have requested that the TENS device, when powered on, display the number of hours of use since it was last serviced, but they might agree that this feature can wait until after the core functions and safety mitigations have been implemented.

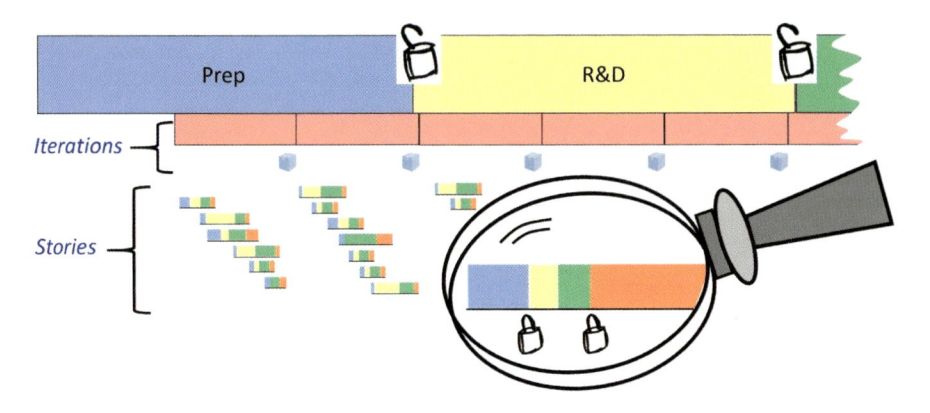

Figure 25 Agile controls development at the individual story level

You get better control over the ongoing work by exercising control at the story level than at the project phase gate level.

So, do we need to have these phases at all? No, but we find them helpful for communicating with the rest of the organization: overcoming organizational resistance, achieving buy-in, and educating people outside development about how the Agile approach meets the intentions of the phase gates. Each of those phases can be described using a variety of nomenclature, each has a meaning that everyone understands, and each stage has a different impact outside of the project:

- Preparation - Project setup, skeleton staff, concept, vision, planning, definition
- R&D - Feasibility period – may not go further. Product is mainly knowledge plus decisions about architecture, design,

market segments, business and regulatory wrinkles, discovery, definition
- Development - Full staffing, engaged with any joint venture or other third parties, DevOps, delivery
- Final test - May be substantial or not apply at all, qualification, independent V&V
- Deploy - We are live with at least some customers, release, launch

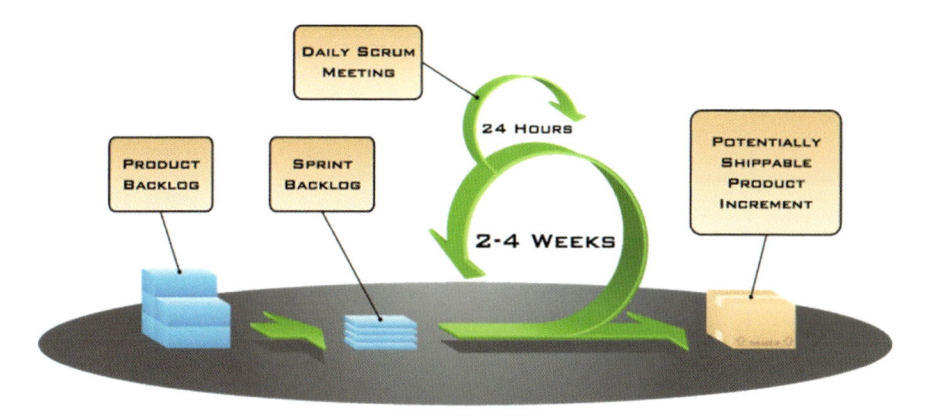

Source: Mountain Goat Software, Creative Commons license CC-BY-2.5

Figure 26 The Agile iterative process, as depicted in Scrum

Iterations Rather than a Linear Schedule

Consider the iterative process outlined in the Scrum method illustrated above.[40]

Scrum is one of several techniques that group under the Agile umbrella; it focuses on project coordination and not on technical practices of development itself. In this process, a product backlog lists all the desired product features, and a sprint backlog lists those features that have been elaborated as stories, which the team can now develop. During an iteration (or sprint), usually two to four weeks in length, the team works only on the stories selected for that time so that a runnable product will result. Each day the team meets briefly in a stand-up meeting to share progress and discuss any blocking issues. Note that along with features, tests for those features are being developed and executed cumulatively as the project goes along.

At the end of each iteration, three events occur:
- Demonstration of the stories completed
- Retrospective
- Planning session for the next iteration

The demo is a chance for the team and stakeholders/customers to see the completed stories (preferably through an actual physical demo of the evolving product), to get a deeper understanding of how they work, and to discuss what now becomes possible to build upon this base. Take our TENS device as an example: After the team implements the start-up sequence and the delivery of pain-relief pulses, stakeholders can try out the system and indicate what sort of information they believe the system should display during therapy. For team members, the demo is a chance to get a better understanding of the wider goals that the stakeholders/customers have in mind. They may suggest a way to achieve those goals that the others would not have known to ask for.

The retrospective is a chance for the team to reflect on what can be improved in their work processes and interactions with each other. Retrospectives are a big topic, more than we can cover here. One of the biggest mistakes a team can make is to do away with them to save time. (Quality assurance will recognize these as the "check" step of the familiar plan-do-check-act cycle. Retrospectives provide a test of how the process and interpersonal relations are working, and are a parallel to how test is the check stage for the product itself.)

The iteration planning session is a working meeting where the team and product owner select a set of stories from the backlog that they will aim to complete in the iteration, which then begins immediately after the session. It's important that the stories have already been seen by the team and that they have had a chance to clarify their understanding of what is involved.

As an aside, the fixed-iteration approach is not a divine imperative. Depending on the project and the team, some groups employ the Kanban process (instead of Scrum). In Kanban, a list of worked-out tasks (features) is always available; each time the team finishes one item, it selects the next one to work on. Their selection is based on relative priority established collaboratively with the stakeholders. Kanban may be more appropriate for some types of projects; a discussion of Kanban vs. iterations is beyond the scope of this book. We will focus on the iterative approach here.

Whether by iterations or Kanban, the discipline is the same: start simple; deliver clean, runnable product at each round; and test each new addition to prevent bugs from building up.

Get to Done with Smaller Batches

There is a temptation to use longer iterations (or sprints) when work is difficult or conditions are uncertain. But working in smaller batches—shorter iterations—is actually a strong way to take back control. There is a skill to breaking the work stream cleanly into small testable chunks (stories). It's also important to define a story's "done" criteria up front and to note whether a story contains work that is not well understood, such as the first time you are using a new operating system, you need to port to new hardware, etc. You may need to mitigate the risk of trouble; one way is simply to do risky stories early in the project so that you keep more options open. Tasks that are not well understood but still must be done are always present in development work, and Agile has many techniques for dealing with them.

Chapter 6 – Applying Agile: Agile Teams and Environments

Agile Project Roles

Within the process we've described are roles (product owner, scrum master, development team), artifacts (product backlog, sprint backlog, increment), and events (sprint, sprint planning, daily scrum, sprint review, and sprint retrospective), to give an example from the Scrum framework.

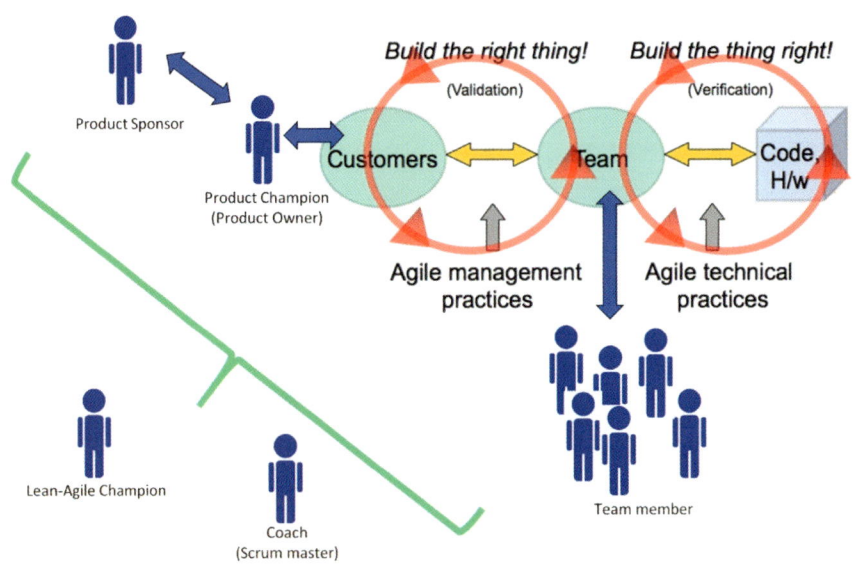

Figure 27 Roles in Agile development

The various roles in Agile development are pictured in Figure 27. The Product Owner role is to optimize value delivered: order the items in the product backlog to best achieve product goals, ensure that the product backlog is visible, transparent, and clear to all; and ensure the development team understands items in the product backlog to the level needed. In our TENS project, for example, the Product Owner would discuss priorities with the stakeholders and ensure that the team addresses the correct pulse algorithm and the need to display the duration of therapy—elements fundamental to the system's function—before implementing other less-crucial self-checks.

The Scrum Master is a shepherd of sorts: one who helps everyone understand and use Scrum; removes impediments to the Development Team's progress; facilitates Scrum events as requested or needed; and helps the Team, PO, and Organization realize the full benefits of Scrum. This person's best tool is influence, rather than positional authority.

(Nancy) The role names can be different when not using Scrum, (e.g. "Coach" instead of "Scrum Master", "Product Champion" instead of "Product Owner") but the duties are similar. Sometimes recognizing additional Agile roles is helpful. In an Agile consultancy where I worked, we found it helpful to identify a role of "Product Sponsor" to be that stakeholder who controls the purse strings of the project. When there are many Agile teams, some organizations choose to have a role that oversees the Agile process at the organization level – Enterprise Coach, Lean-Agile Champion, or similar titles are used.

The Agile team needs to be self-organizing and self-directed: accountable only as a whole team; cross-functional, with all the needed skills; and able to decide how to turn Product Backlog into increments of potentially releasable functionality.

Team autonomy – the condition of being a "self-organizing" group – is a cardinal quality for Agile teams no matter which specific methodology or framework within the Agile umbrella they use. The idea is that for technical work, especially in development, the expertise rests primarily with those in the teams not their managers. The team is best positioned to decide **how** the features should be created, while the decisions on **what** to create come through the Product Owner. Self-organizing does not mean that team members must assign themselves to teams, although some do. It means that they are more self-governing. They have more latitude than traditional teams in choosing the technical practices and tools they'll use. [41]

Team Decision-Making

It follows that Agile teams need to have a way to make decisions that all can support, and to be able to revisit them when necessary. An Agile coach should help teams to talk about the way they want to work together and what guiding values they share. Having done this early on helps greatly when strong differences arise – rather than debate endlessly, Agile teams learn to put the energy into creating an experiment, and then all can unify behind doing that. Then in the light of new experience, they can revisit the issue.

Management of Agile teams needs to be more facilitative, and less interventionist as the team becomes able to handle more of their decisions. It's good to work with a coach to clarify which decisions the team should control. For instance, the use of a particular software lifecycle management tool may be a company standard that has to be lived with, but the team may be free to choose plug-ins they prefer. In time the team will be ready to handle more substantive decisions.

Protect the Habitat

Agile teams also need to be supported by providing them with a safe habitat: test environments; access to stakeholders and end users; team room; and intangibles like allowing them to apply themselves fully to one project, having HR policies that don't force-rank people, and having respect for team autonomy.

No Multitasking Allowed

Is it possible to blend high tech skills on a project without losses due to interruptions and task switching? Moving specialists from project-to-project as their skill is needed is widely practiced, but new research has shown surprising downsides to this. Researcher Gloria Mark, Professor in the Department of Informatics at the University of California, said that high tech office workers stay on a task for an average of 3:05 minutes, and when interrupted, it takes 23:15 minutes to resume their work.[42] But some interruptions don't have this disruptive effect; interruptions can actually be beneficial if:

- They are about the same task one is working on
- A problem that someone is stumped on is allowed to "incubate," and then the answer comes to them in the middle of other work

Agile teams are, in theory, focused on one project to completion. With a mixed skill-set selected to match the work stream, interactions lead to the positive kind of interruption. When a company keeps teams together on one project, they avoid the tremendous waste that occurs when individuals are shifting on and off several projects, trying to apply their skills only where needed.

People do not get better at multitasking with practice. Research has shown that this is not like other skills. The ability to multitask stems from the "executive" part of the brain, the prefrontal cortex.[43] This area is the part of the human brain that is most damaged as a result of prolonged stress, particularly the kind of stress that makes a person feel out-of-control and helpless. The kind of stress, say, that you might feel when overwhelmed by the demands of multitasking. The hippocampus—critical to the formation of new memories—is also damaged by this kind of stress. The more you multitask, the *worse* you become at it.

Multitasking is for operating systems, not people.

If a government told business that every project they operate will now be taxed at 40 percent, can you imagine what would happen! Time-slicing their people over three projects results in 40 percent of the people's time being wasted.

Number of Simultaneous Projects	% Time Available per Project	Loss to Context Switching
1	100	0
2	40	20
3	20	40
4	10	60
5	5	75

Figure 28 Time lost to context switching.[44]

Time-slicing people to cover several projects is at epidemic levels. But it cannot work; no one can context-switch with zero time lost. Having people 100 percent dedicated to a single project seems wasteful only because companies are organized in such a way that they cannot maintain the flow of information, decisions, and tool availability that is necessary to keep dedicated people productively utilized.

That is exactly the goal a would-be Agile organization has to adopt.

Chapter 7 – Applying Agile: Cumulative Documentation and Risk Management

Document as You Go

Stories are what you agree to build. The regulatory bodies take no position on the use of Agile or any other methodology. The FDA's General Principles of Software Validation refers to "...predetermined and documented software requirements" and "...predetermined criteria for acceptance of the software."[45] No standard or body explicitly states a required format for your requirements—refined and captured, and with conditions of satisfaction defined, your stories are requirements by any other name. For an Agile team's documentation to meet regulatory requirements, it needs to conform to the company's stated policies. Their policy can be that Agile user stories with acceptance criteria are an allowable format for stating what the product will do. Within each story, defining the conditions of satisfaction establishes your tests—and the traceability between tests and requirements. If your Agile team accumulates its refined user stories, its design information, and its tests as development proceeds, then when the product is ready, the documentation is ready as well! Keep the process straightforward, so you can document at the time of generation. Use your tools if you can (e.g., story tracker and automated test harness) to keep the process seamless, and encourage your quality assurance colleagues to help design and manage the system. QA will need to sign off that you've followed a process, so why not seek their suggestions as early as possible?

The diagram in Figure 29 illustrates the cumulative documentation approach. At the left is the preliminary set of features the team sets out to build; as the team refines them into finished stories, they are added to a growing Software Requirements Specification (SRS). Each of the circular arrows represents an iteration/sprint, in which stories are implemented; design information is added to the software design specification (SDS); and tests for the implemented stories are written, executed, and added to a growing set for verification and validation (V&V). When the product emerges, all of these artifacts (requirements document, design document, and verification / validation tests) will be ready as well. (Not shown in this diagram is traceability—that information can be generated cumulatively along with the rest of the documentation.)

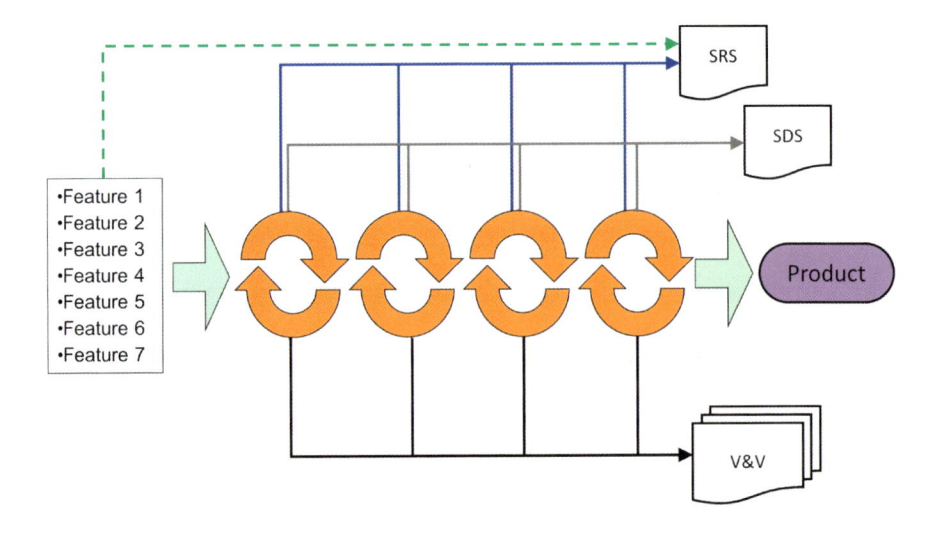

Figure 29 Documentation can be cumulative

To make this kind of cumulative documentation possible, it may be necessary to set up more than one tool, or at least to establish certain disciplines in using the tools available. Some story tracking tools are only designed to track stories, and provide no inherent linkage to, say, test tools (for traceability from requirements to tests). In these systems, referencing the corresponding test or tests in a story description establishes traceability. Some systems, like JIRA, provide the ability to link stories to tests, for example, but require purchasing additional modules and going through a non-trivial setup process. In some cases, it may work best simply to track information in a spreadsheet, outside the story tracking and test tools.[m] No one approach is best in all cases. The point is to record and track information as you generate it and not leave documentation for "later."

Product demonstrations are also a great opportunity to make the Agile process work for you. If you meet with customers and

---

[m] Make sure to validate the tools.

stakeholders to go over current development and obtain feedback, you can easily record the meeting—including attendees, features reviewed, and action items—in a memo to file. The series of those meetings are incremental design reviews; why not take credit for them? Then, when you hold a complete, final design review with all stakeholders, you should have no surprises.

When to Establish Change Control

Teams often ask when change control begins in Agile approaches. Simply stated, you need to institute change control once you've exited the research/feasibility stage and embarked on true development. The rest of the question, however, is about which items you should exercise change control over and how you should implement change control.

If some features require more exploration (research) than others, then development on those features, and therefore change control for the information, starts later.

The development documentation you need to create is the same as it always was: requirements, hazard analysis, design, traceability, verification/validation tests, and results.

Handling change control as suggested ***does*** require you to describe how it will be done. At first this might be your development plan; eventually, if you've settled on your tools and approach, you could write it up as an SOP.

Reviewing Information Directly in the Knowledge Tool

(Brian) Sometimes the question comes up of whether JIRA or some other knowledge-capture tool can be reviewed instead of paper documents. I can only give you my own view on this question; you may find inspectors or notified body auditors who disagree with me.

Everything depends on how your tool is set up and used. Can you configure the tool so that once a user story is refined, it can be locked and electronically approved? You may not need many approvals, so keep the process light and simple. If you can review stories as recorded and approved (and if you have validated the tool for electronic signatures, per 21 CFR Part 11), then you can review the stories and their tests individually or in small groups, rather than as massive, printed-out (or exported) documents.

The short answer is—with electronic signatures and control of editing privileges, I believe you can justify reviewing the information you've captured in your tool. Just make sure you've explained how you exercise the equivalent of document control in your development plan.

Documenting Sprint Demos as Mini-Design Reviews

(Brian) I don't know of any established precedent for documenting your sprint demos, though when I asked a colleague who has long used Agile methods in medical device development about sprint demos as mini-design reviews, he emphatically agreed with me.

My suggestion is to keep it simple. Use whatever internal memo format you already have in your organization.

- Have the memo be "To: file" and give the date of the demonstration
- List all the participants in the demonstration (include their functional groups or responsibilities—quality assurance, regulatory affairs, marketing, product management, clinical affairs, or whatever)
- Describe what was demonstrated and discussed (i.e., if specific items besides product features were covered)
- List any issues identified, such as features that need to be modified or hazards that were identified
- Note anything else significant I may not have mentioned here
- List all action items

The point is to keep this as simple as possible, so that you can take credit for an activity you're doing anyway.

In 21 CFR 820, the FDA is purposely vague about design review. So many types of product development take place that no one set of reviews would fit all cases. All the FDA says is:

> *Design review. Each manufacturer shall establish and maintain procedures to ensure that formal documented reviews of the design results are planned and conducted at appropriate stages of the device's design development. The procedures shall ensure that participants at each design review include representatives of all functions concerned with the design stage being reviewed and an individual(s) who does not have direct responsibility for the design stage being reviewed, as well as any specialists needed. The results of a design review, including identification of the design, the date, and the individual(s) performing the review, shall be documented in the design history file (the DHF).*[n]

The section for "Design and Development Review" in ISO 13485 is worded very similarly—it requires the company to have "planned and documented arrangements" to evaluate how the design meets its requirements and to propose necessary actions.

That said, it's common for reviews to be conducted on design inputs, on verification, and on validation. Those points make sense in the flow of a development project. However, a company can define its reviews differently, in ways that make sense for its approach to development. Here's a possible set of reviews:

(a) Market opportunity and technical feasibility (prior to design)
(b) Initial backlog established (high-level design input)
(c) Design sufficient—when it appears that adequate product design has been created, management can review the STORIES and hazard analysis (detailed design input)
(d) Verification and validation complete (design outputs)

Items (c) and (d) can be triggered at any time during a routine iteration demo, and a full, formal meeting called at that point.

---

[n] The FDA Design Control Guidance describes reviewing all design inputs before implementation, but also discusses reviews when using concurrent engineering (a concept which the Agile approach embodies).

As to who needs to participate, both the FDA and ISO requirements are clearer. At a full, formal design review, *all functions* concerned with the design—engineering, marketing, clinical affairs (if you have such), service (if appropriate), management, and so on—need to take part, with at least one person not directly responsible for the design taking part.

If you document your sprint demos as "mini design reviews," these are informal and don't need to have the full complement of participants that you need for a full design review. The point of documenting these as mini design reviews is to arrive at the full design review with few to no surprises.

(Brian) Neither the FDA nor ISO lists specific documents that need to be reviewed. Determining what documents are relevant and necessary is up to your organization. However, if I were planning for design reviews or acting as inspector, I would want to know several things:

- That sufficient inputs (for software stories and for other areas, whatever you call the inputs) have been captured and understood
- That a responsible extent of hazard analysis has been performed, and hazard mitigations are linked to stories for software, or as appropriate for other areas of design
- That all stories trace to verification tests, and that the verification tests have been executed, reviewed, and passed
- For a sufficiently high-risk product, that stories are traced to where they're implemented in the design

Safety—An Emergent Property

A number of techniques help in managing safety risk, but you need to avoid tunnel vision when assessing how injury or death could occur. We can get a good idea of potential problems in our analysis, but we can't predict all the ways a product could fail. (If we could, we'd design and build to prevent any issues in the first place.) Safety must be seen as an emergent property. No one component of a system causes an unsafe condition. Rather, a series of events and malfunctions are needed. Building a safe system (i.e., medical device) requires understanding how the entire system can behave and designing it in a way that avoids unsafe behavior.

Inevitably, we will understand more about potentially unsafe product behavior as the design takes shape, as users try out the design, and as more stakeholders are brought in to see our progress.

Effective risk management hasn't changed in the Agile environment; it's still a plan-do-check-act process, and it still lasts through the entire product lifecycle. The benefit of Agile is that, being an iterative learning environment, we have many more opportunities to recognize unsafe situations and find ways to mitigate them. In our iterations, we need to stay alert to safety questions and make sure to brainstorm those as part of our iterative design reviews.

Discussing and Documenting Risk

(Brian) There is also no single correct answer for how often to discuss and review safety risks. If your product is inherently higher risk—say, it delivers energy such as x-rays, delivers or removes fluid from the body (especially something that could be fatal if overdosed), or supports a life-critical function (like a heart-assist pump device)—then the risk discussion should take place as early and as often as possible. I'd recommend re-examining the hazard analysis every iteration (sprint).

If your product is relatively low risk—that is, direct harm to the patient, caregiver, bystanders, or environment is highly unlikely (for example, an automated blood pressure measurement device)—then it may be sufficient to ask, at every product demo, whether the participants are aware of any new hazards or any changes to severity of hazards already recorded.

When and how often to revisit the hazard analysis depends entirely on the type of product your team is developing.

The common element is not the question of how often to revisit the risk analysis, but the requirement that you keep track of all the hazards you identify and trace their mitigations to something in the requirements or design. Some mitigations may not be in software (such as protecting against overcurrent via a circuit breaker), and wouldn't trace to software stories. But those handled in software should always trace to corresponding risk mitigation stories. Your ongoing documentation can ease this process.

Chapter 8 – So (We've Sold You on the Benefit and) You've Decided to Become Agile

To get your Agile program started, you need several key elements:
- Stakeholder buy-in
- A pilot project
- A coach
- A solid plan for communications

Buy-in from all your stakeholders is such an obvious need, but it's one that is easily glossed over. Management needs to be willing to accept that this is a viable method, as well as accepting that Agile generally favors broad initial estimates of how long a project will take. Estimates will improve as the team gains experience. Marketing needs to be willing to take part in frequent demonstrations and accept trade-offs when they arise. Engineering needs to be ready to invest in learning a new way of working, rather than just trying this Agile thing because someone says they have to. You should also try to determine whether fear is holding anyone back. Does *anyone* in the company fear that his or her job will be eliminated if Agile takes hold?

Buy-in is easier to obtain, and success is much more likely, if an appropriate pilot project is selected. A pilot project will allow your team to take baby steps, to experiment and see what works for them. Trying to take all projects to an Agile approach at the same time is extremely likely to crash and burn, so unless your development group is very small, don't do that!

Though it may seem less expensive to introduce Agile from within than to employ a coach, you're likely to waste lots of time and effort. The guidance of a coach will help you anticipate those hiccups and avoid the blind alleys.

Selecting a Pilot Project: Can This Project Go Agile?

Managers often ask whether a given project is a good candidate for the use of Agile software development practices. The answer depends on a number of factors:
- Would this be the first Agile project for this company?

- What is the nature of the project's funding?
- How important is the project to the business?
- What external dependencies does the project have?

Another good question to consider is whether Agile practices are necessary for a given project. Agile handles complexity well, and for truly complex projects it may be the only viable method. Uncertainty can exist in the requirements area and also in the technology realm.

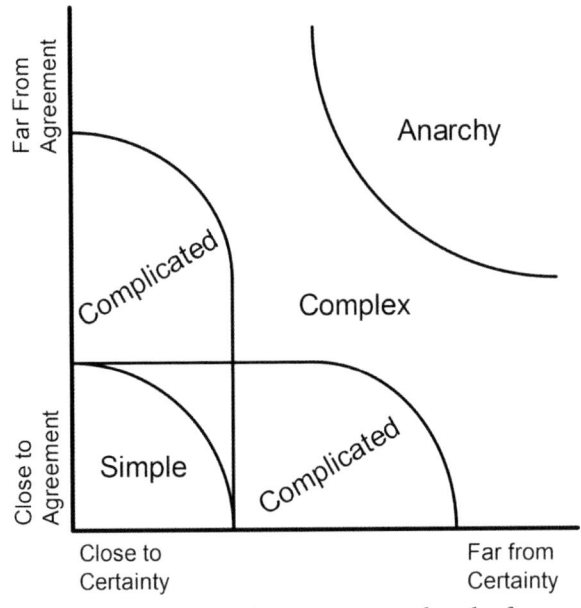

Figure 30 Project technical certainty vs. level of agreement on objectives.[46]

Figure 30 gives an idea of these two uncertainties in action. Some projects have enough uncertainty that they are not plannable in the waterfall sense. Those must be done using Agile if they are to succeed, but the company may not be ready to support Agile practices. In that case, the project is simply not within reach for that organization.

(Nancy) I always ask "Would this be the initial introduction of Agile for this company?" If yes (or Agile is still new to the company), then I am more stringent on lining up the success factors. The success factors are listed below in order of importance, along with some discussion of each.

Success Factors

Is this project bringing in new revenue? If so, then it may be able to pay its way, which will often allay the criticism that having staff dedicated 100 percent to one project is wasteful. Projects paid for as overhead (e.g., regulatory compliance, infrastructure) are less likely to be given the high priority that an early Agile project needs. They also do not usually have clear, urgent sponsorship.

Is there someone on the business side who really wants this to proceed? Agile projects need sustained levels of involvement by a core team member called the Product Owner—someone representing the needs of the customer and who can work frequently with the developers. (In some cases, a Product Owner team is best.) This is done *instead of* investing labor to create a huge specification document. Clear sponsorship by someone willing to spend substantial amounts of time with the team is a necessity. This person should be able to answer, on the spot, 80 percent of the questions the team asks without being overruled. This is necessary for team speed.

Does the project have many external dependencies? If the project has tight dependencies on other projects, external vendors, or a distributed team, it will be harder for the team to achieve the speed that Agile is known for. If you can't get the speed, it's more difficult for the business to see enough benefit to want to sustain an Agile team.

Will there be sufficient time and resources to put together a strategy and plan for the team? Strategic planning for the project as well as planning for Agile training and coaching requires commitment from a senior leader who will be one of the Agile Champions going forward. Success criteria will be established as part of the plan and regular conversations scheduled with the Agile Champion.

Will the company allow the core team to focus full time on the project? It is vital that Agile teams stay focused on their work. Switching between projects and dealing with work handoffs is very disruptive to the kind of focus that is needed for speed and quality. If some members are three-quarters time, that can be workable as long as other work is not higher priority than the Agile project.

Will the company provide a team room? (Nancy) Co-location of the team is fundamental to speeding up interpersonal communication, and communication is the lifeblood of Agile teams. Having coached dozens of teams, I have seen co-located teams progress much faster than others just from the advantage of easy communications. (There are some distributed teams that work 100 percent of the time by video link, and this can be viable for moving to Agile. Ease of communication is key.)

Will the team be able to deliver work to real users periodically? Agile teams need the feedback that can only come from delivering software to real users. They need to do this at least every couple of months. If no user feedback can be obtained for long periods, that's a danger sign. It's too easy for the team to deliver the wrong features in the absence of real feedback. Is it possible to negotiate with your customer to review features as they are completed? The customer should also be available to address questions.

Can the team have access to all the tools necessary to fully test their code incrementally? There is a huge emphasis on testing in Agile teams, and it's vital that they work the highest-priority features to completion in case a scope cut becomes necessary. Plan to create unit tests and be able to automate many of them.

Can the requirements remain negotiable throughout development? If conditions are such that all the team's deliverables have been defined and there is no flexibility allowed, then the project is not suited to Agile. Too often, "negotiable" is interpreted by stakeholders to mean that features will be dropped. High-priority features can be very reliably implemented, but only if there is willingness to cut scope for less important ones. Adjustment of scope is the main tool that Agile teams use to deliver regularly. If an organization regards every feature as "top priority," then negotiability is not present, and the team will have a difficult time controlling their work.

Does the team have coaching support? If this effort is among the first to use Agile practices, everyone involved with the project should get the appropriate type of training followed by regular support from a qualified Agile coach.

Getting to "Now"

Once you've considered these success factors and selected your pilot project, you'll need to get to "now" (where you're ready to begin iterations). This typically requires assembling your team, preparing several iterations worth of stories, selecting tools (automated test harness, code repository, bug tracker, story tracker), and doing some training on Agile testing. Remember, don't try to put too much detail in milestones that are further out in time.

An early iteration, comparable to the more traditional prep and R&D stages, can be used to explore your options for approach and tools before launching out in earnest. We'll talk more about a potential "Iteration Zero" a little later.

(Nancy) I worked with an industrial controls company to help them decide how to start a huge internal project that would be their first use of Agile methods. After we re-cast enough of their traditional style specs into several iterations' worth of Agile stories, and laid out major milestones for the initial 6 months, it was time to ask them the big question: pointing to the timeline chart where "Now" was written on the point where the first iteration began, I asked "Can this be tomorrow?"

Immediately everyone knew many reasons why not. So we made a list, and before long it was clear that there were two types of things on this list: work to do, and decisions to make. I next asked them to arrange the items so we could tell which decisions enabled which work items to proceed. We needed to know who could make the decisions. Very soon it started looking like the plan for an iteration to bootstrap everything. That's how they got started. Their CTO took up the slogan "let's get to Now!" and drove those decisions to completion.

A solid plan for Communications

In place of heavy up-front planning and complicated stage-gate procedures, Agile companies use a lot more instances of ongoing communications. Here are some examples:

- Team members will synch with each other daily, and this is kept brief by use of a well-maintained story board (or Kanban board) – a communications device.
- Team members keep their energy focused on the job at hand because they have already made a set of "working agreements" they will use for self-governance. They will revise these any time they see a need.
- Few Agile teams can be 100% self-sufficient. Managers arrange for a bit of time from others outside the team to help with specific stories if needed. Part of iteration planning is these negotiations with other managers that happen before an iteration starts.
- During an iteration all the status information is readily visible in the team room or a dedicated space. Story board, Risks & Issues board, Team Blockages list will all be visible for anyone to see. Product owner and stakeholders should watch these for signals so they can clear or prevent blockages that hinder the team.
- The Iteration Review (also called "Demo") meeting is a great opportunity for team and stakeholders and customers to learn where the product stands now, and what things are possible now.
- There should be two kinds of retrospective – one just for the team members to discuss how to improve their interactions, and issues within their control. A second type should be for reflecting on issues that go beyond the team's control – to identify patterns of blockages that occur, and any other issues that are outside the team's control to resolve.
- Story grooming meetings should occur regularly at a pace sufficient to keep enough well-understood stories ready for being slotted into an iteration. In this case "ready" has a very specific meaning: There is an agreed "acceptance criteria" for the story, and the team has no further open questions about what it entails.

Once an iteration is started, there is a need to steer by the information that becomes available each day, and if reality starts to diverge too far from the expectations, it's better to cut some scope, if necessary. The ongoing communications described here are very different and much more interactive compared with traditional development practices. It's important for product owner, team, and key stakeholders to consider how to support these increased communications because they are vital to the success of Agile; they do the work "live" that the heavy process did up-front.

Chapter 9 – Agile Planning Techniques

Traditional managers and quality specialists often balk at the fourth item in the Agile Manifesto, which states that Agile values responding to change over following a plan. "What?" they cry. "You mean you Agile folks don't plan?"

Nothing is further than the truth. Planning occurs in Agile at many levels, from the overall project to epics to sprints to daily activities.

In this chapter, we describe two specific techniques not central to the Agile approach but extremely helpful in safety-critical work. The first, Impact Mapping, addresses the question of what product to build. The second, Story Mapping, sets out how to build the target product. We'll return to our TENS example to demonstrate these techniques in action.

Impact Mapping—Envisioning Your Product

How do we envision a new product? How can we improve the marketing planning so that multiple points of view can contribute?

Gojko Adzic has supplied a highly useful approach in his book *Impact Mapping*,[47] a technique that aligns product rationale with the company's goals, maps business value creation in a way that multiple functions can understand and contribute, and accommodates changes.

The keys of impact mapping are actually simple:
- Why: Why are we doing this?
- Who: Whose behavior do we want to impact?
- How: How should our actors' behavior change?
- What: What can we do to support the required impacts?

Analyzing the TENS example permits building a sample impact map, which shows and connects your goal, your actors, your desired impacts, and your deliverables.

The goal—the Why—is to sell a specific number of units in a specific time (shown as 2000 units in the first three years).

The types of people you need to impact—the Actors—becomes clear in light of the intended product use.

- Physicians who will operate the instrument
- Patients who are experiencing chronic pain
- Regulators who will review the product for clearance to market
- Support personnel who will need to service the instrument.

The impacts you need to achieve are specific to each of the actors:

- Physicians will want confidence that they can adopt TENS in their practice with little to no training
- Patients will want two impacts: assurance that they will not be injured, and reduction or elimination of pain
- Regulators need assurance that the system is effective and safe if used properly (and that "reasonably foreseeable misuse" is largely prevented)
- Support will want you to make their jobs easier

Finally, the deliverables—the features we aim to create—follow from the kinds of impacts we aim to have. These are shown in the sample impact map:

- A prompted setup sequence, to keep operation simple for the physician
- Limits on the settings and fault detection during therapy, to assure patients of their safety
- A proprietary pulse algorithm (the result of much previous research) for highly effective pain relief
- A convincing safety profile

The TENS impact map is a simplified example, but it shows the key elements of the method. We won't dive deeper into this here, but Gojko Adzic's book can help you understand what questions to ask.

Impact map figure

Goal: Sell 2000 units in first 3 years

Actors: Physicians, Patients, FDA / regulatory body, Support

Impact: Can adopt TENS with confidence; No concern about injury; Pain is eliminated or reduced; Grant clearance; (other)

Deliverables: Prompted setup sequence; Settings Limits; Fault Detection during therapy; Proprietary Pulse Algorithm; Convincing safety profile

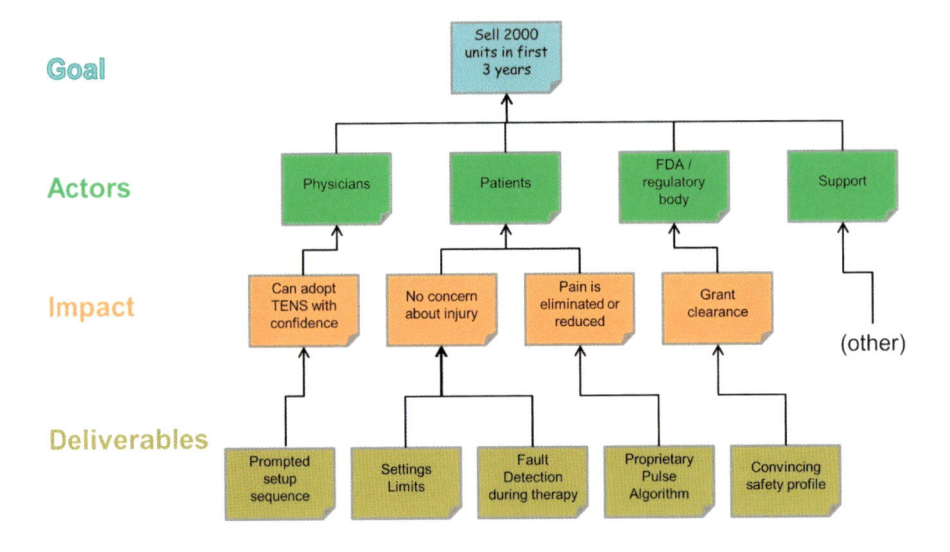

Figure 31 Impact map for the TENS project

Laying out the impact map in this way allows people with different types of expertise to understand and contribute. Clinical specialists, service personnel, marketing experts, and engineers can all review the map and point out whether it misses some important element. An impact map therefore helps align stakeholders, but it also keeps your aim on a moving target—market demands can change quickly. Finally, the map focuses on behavior: the behaviors that will support or block the product's success.

Story Mapping—Bringing the Product to Life

Once you've planned what product you will develop based on whom you aim to impact with the product, it's time to create an outline for how you intend to bring that product into existence. For this process of planning feature sets and releases, the Story Mapping technique is extremely useful—project managers, marketing, and hands-on technical teams can all participate in the mapping process. Jeff Patton describes the technique very well in his book *User Story Mapping*.[39]

Start by creating a horizontal axis that consists of the sequence of activities you envision for a customer using your product. The

Agile Methods for Safety-Critical Systems

vertical axis (from top down) defines increasing levels of support for a particular step in the user journey. This approach avoids releases that are unusable because they depend on less urgent stories that have not yet been implemented. In the context of a medical device, once you know the sequence of user events, the levels could be laboratory testing, feasibility study, and pivotal clinical study versions. The group of features lined up under a given product-use action all support that specific step.

The TENS example shows how the story map works. Across the top is the sequence in which the caregiver sets up, runs, and shuts down the device for a therapy session. Under the steps of that sequence are features that support the corresponding action—for example, where the action is to enter treatment settings, the supporting feature is for the system to prompt for treatment duration (and display this throughout treatment).

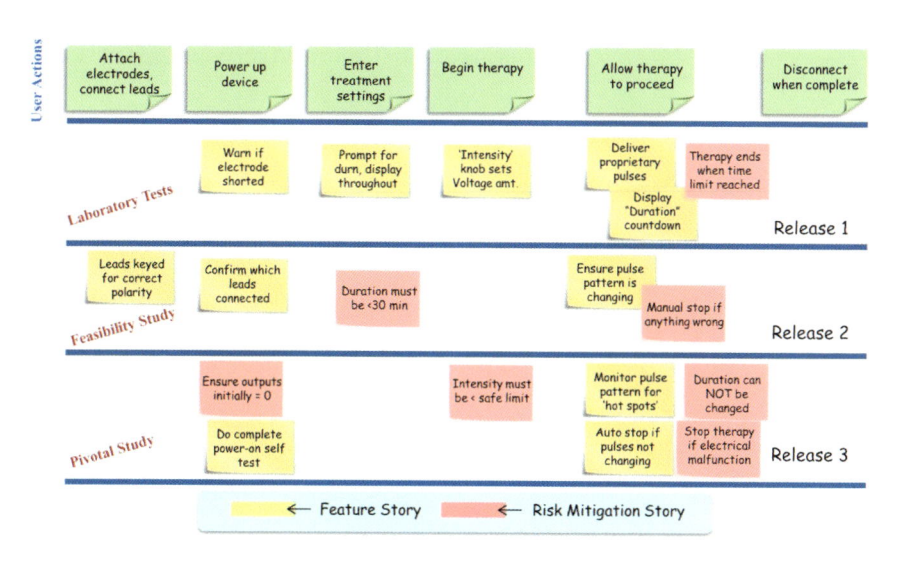

Figure 32 Story map for developing the TENS system

Some of the features in the map may be mechanical or electrical rather than software—this is fine! In the example, the first step is to attach electrodes and connect the leads to the instrument. The supporting feature "Leads keyed for correct polarity" is purely mechanical.

Notice that some stories in the map are denoted as risk mitigations. These trace back to hazards identified in some previous analysis, but the risk management process doesn't stop there. As you add to the map, additional hazards will become clear, and you'll add other risk mitigation stories where needed. If marketing demands some new feature in the middle of development, you can evaluate hazards related to that feature and add both the new feature and its risk mitigations to the map. This fits with the "ongoing risk-evaluation process" of ISO 14971.

If you use a software tool[o] (for instance, TechTalk SpecLog[p]) to build the story map, you can also easily generate the traceability necessary for documentation: from stories up to user actions, from stories down to corresponding tests, and much more.

Table II shows an example, generated from SpecLog, of a combined impact map/story map exported in table form.

The Business Goal, Actor, Impact, and Deliverable rows constitute the Impact Map for this product. Note that deliverable D47 was added much later—it was recognized later that this deliverable would be vital to the product. The impact mapping approach readily accommodates additions of this sort.

Each element with an ID (A3, D15, UA16, RM24, US27, and so on) is a "card" on the SpecLog workspace, which looks similar to Figure 32 above. Items labeled "US" are user stories, and those labeled "RM" are risk management stories, which were derived from a separate hazard analysis. Both the US and the RM items correspond with the traditional concept of "requirements" but are developed as user stories. Each item is connected to one or more parent items (which establishes traceability). A complete user story or risk story is the combination of actor, goal, and value ("As an [actor] I want to [goal] so that [value]"). "AC" is an acceptance criterion—the AC

---

[o] Tools must be validated for intended use.

[p] Mention of specific tools for Agile project management, software development, or related work is used in this book to give examples with detail and clarity. The authors have no business or financial connection to any of the vendors of the tools mentioned.

title and AC details define a test that will be executed, and the acceptance criterion (shown in italics in this table) appears just below the corresponding story, which creates story–test traceability. Not every story in this example has defined acceptance criteria, though in practice, every story would include at least one acceptance criterion, and many would include several.

Table II, Export of Impact Map/Story Map Contents

ID	Type	Actor	Goal	Value	AC Title	AC Details	Parents
BG1	Business Goal		Sell 2000 units in the first three years				
A2	Actor	Physician / caregiver					BG1
A3	Actor	Patient					BG1
A4	Actor	FDA/regulatory body					BG1
A5	Actor	Service Engineer					BG1
I6	Impact		Can adopt TENS with confidence				A2
I7	Impact		No concern about injury				A3
I8	Impact		Pain is eliminated or reduced				A3
I9	Impact		Grant clearance				A4
I10	Impact		Calibration / adjustment is possible				A5
D11	Deliverable		Prompted setup sequence				I6
D12	Deliverable		Settings limits				I6
D13	Deliverable		Fault detection during therapy				I7
D14	Deliverable		Proprietary pulse algorithm implemented				I8
D15	Deliverable		Separate adjustment / servicing routine				I10
D47	Deliverable		Convincing safety profile				I9

Agile Methods for Safety-Critical Systems

ID	Type	Actor	Goal	Value	AC Title	AC Details	Parents
UA16	User Activity	Physician / caregiver	Attach electrodes, connect leads				
UA17	User Activity	Physician / caregiver	Power up device				D11
UA18	User Activity	Physician / caregiver	Enter treatment settings				D11, D12
UA19	User Activity	Physician / caregiver	Begin therapy				
UA20	User Activity	Physician / caregiver Patient	Allow therapy to proceed				D13, D14
UA21	User Activity	Physician / caregiver	Disconnect electrodes and leads when complete				
RM22	Risk Story	Physician / caregiver	Ensure that outputs are initially set to zero	Patient will not be burned			UA17
					Zero prompt	Given the electrodes are applied and leads are connected- When the device is powered up- The software shall prompt to zero all channel intensity controls.	
RM23	Risk Story	Physician / caregiver	Treatment duration cannot be set >30 minutes (FTA-2)	Patient will not be harmed			UA18
RM24	Risk Story	Physician / caregiver	Ensure that intensity cannot be set higher than safe (FTA-4)	Patient will not be harmed			UA19
US25	User Story	Physician / caregiver	See that the pulse pattern is changing	I know the pain remedy is being administered			UA20
					Progress indicator	Given that the unit is powered up and connected- When voltage pulses are being applied to the electrodes- The unit shall display an indication that this is occurring.	

Agile Methods for Safety-Critical Systems

ID	Type	Actor	Goal	Value	AC Title	AC Details	Parents
					Pulse display	*Given that the unit is powered up and connected- When therapy is being administered- The unit shall display which pulse pattern is being administered.*	
RM26	Risk Story	Physician / caregiver Patient	Ensure that therapy will stop on electrical malfunction (FTA-1)	Patient will not be harmed			UA20
					Impedance response	*Given that therapy is occurring- When sensor circuits detect excessive impedance- Then therapy shall be stopped.*	
					Short circuit response	*Given that therapy is occurring- When sensor circuits detect a short circuit- Then therapy shall be stopped.*	
					Open circuit response	*Given that therapy is occurring- When sensor circuits detect an open condition (electrode disconnected or wire broken)- Then therapy shall be stopped.*	
US27	User Story	Patient,	Have unit deliver pulses by the proprietary algorithm	My pain will be relieved			UA20
RM28	Risk Story	Physician / caregiver Patient	Prevent increasing therapy duration once it has started (FTA-3)	Patient will not be harmed			UA20

ID	Type	Actor	Goal	Value	AC Title	AC Details	Parents
RM29	Risk Story	Physician / caregiver Patient,	Ensure that therapy stops when stop time is reached (FTA-2)	Patient will not be harmed			UA20
US30	User Story	Physician / caregiver	Set the voltage via the Intensity knob	The therapy will begin			UA19
RM31	Risk Story	Physician / caregiver	Stop therapy manually if something appears wrong (FTA-1)	Patient will not be harmed			UA20
					Manual stop	Given that therapy is occurring- And something appears wrong- When I give the STOP command- Then therapy must stop immediately	
RM32	Risk Story	Caregiver	Have the unit warn if an electrode is shorted	I will avoid shocking or burning the patient			UA17, RM33
US36	User Story	Physician / caregiver	Have the unit conduct power-on self-test	I can be confident the therapy will proceed without problems			UA17
US37	User Story	Physician / caregiver	Have the unit prompt for duration and display throughout therapy	I can be confident the therapy is proceeding correctly			UA18
US38	User Story	Physician / caregiver	Have electrode leads keyed to assure correct polarity	I will minimize the chance of harming the patient mistakenly			UA16
US39	User Story	Physician / caregiver	Have the unit confirm that electrode leads are connected	I know the system is correctly set up			UA17
US41	User Story	Physician / caregiver	Have the system display a duration countdown	I can manage workflow around patient therapy			UA20

ID	Type	Actor	Goal	Value	AC Title	AC Details	Parents
US42	User Story	Physician / caregiver	Monitor the pulse pattern for "hot spots"	I know the therapy is correctly randomizing the pulses delivered			UA20
US43	User Story	Physician / caregiver	Have the unit stop automatically if the pulse pattern is not changing	I can be confident the correct therapy is being administered			UA20

Metrics – The Good, the Bad, and the Ugly

Once you've mapped out your project, including the goals and values, consider how you will measure whether your goals are being achieved. Recall that the point is to provide transparency, so that all stakeholders can understand how development is moving forward, and so that all team members can see the effects of the techniques they are using. Some metrics support these purposes and some don't.

Metrics involve the concepts of objectives, outputs, and outcomes. Our *objectives* are the goals we set (such as ensuring that all registered cars are roadworthy). Our *outputs* are the specific results of actions (fewer cars with mechanical faults), and the *outcomes* are what we are hoping to achieve (increased road safety).[48]

In our TENS example, one *outcome* we aim for is to instill confidence in physicians, patients, and regulators that the therapy is safe. The corresponding *objectives* would be to design the system so that it is inherently difficult to shock or burn a patient. The *outputs* (deliverables) would then be maximum settings, time limits on therapy, and warnings to report shorts or open circuits during therapy.

The delivery team is responsible for outputs, but the outcome for the customer is the responsibility of the project stakeholders. It is the stakeholders (acting through the Product Owner) who choose and prioritize the product features to be built.

The most useful metrics are focused on outcomes (Were the TENS physicians convinced our product is safe?) rather than on outputs (Did the team complete more than twenty-five story points of work in sprint number five?). After all, getting the desired outcomes is the real test of product success.

This concept of "objectives, outputs, and outcomes" is based on the work of Osborne and Gaebler as described by Rob Thomsett in *Radical Project Management*,[48] and we suggest this book for more information on the subject. We'll discuss metrics more thoroughly in the next chapter.

Chapter 10 – Tracking Progress and Accelerating Learning

Agile projects typically can't be tracked the same way that traditional projects are tracked. Think back to Chapter 5: we're not using phase gates. At the same time, by delivering early and often, we are always producing a useful end product (minimum marketable feature set, or MMF). However, the business side of your organization is going to want to know how far along your project team is, how they're spending their money, etc. So how do you measure progress in Agile? It requires a new approach to cost, time, and completion estimation.

When Do You Know for Certain How Much a Project Costs?

Unlike retail transactions—getting a haircut, for instance—our knowledge of the final cost of a project is theoretical for much of the course of the project, which represents an economic risk to the company. Agile methods can help us to lessen that risk greatly. At every iteration, we aim to have a potentially shippable product. (That's not usually possible for embedded systems after the very first iteration. See the section on Iteration Zero later in this chapter for more on this.)

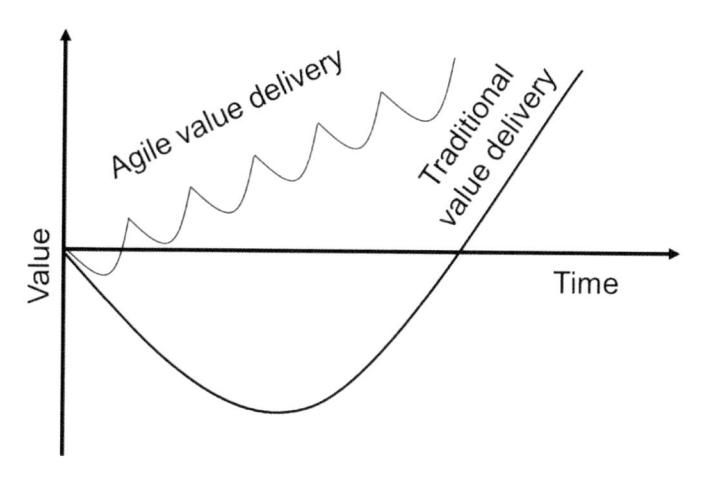

Figure 33 Value Delivery over time

You can design milestones into the plan that can potentially become early termination points with customer value having been delivered. (We mentioned just such an example in Chapter 3.) Hardware lead

Agile Methods for Safety-Critical Systems

times limit the flexibility, but that's why it's common to build a hardware platform that can be improved at longer intervals. Smartphones are a good example of this strategy. This is all part of the strategy of developing the product in vertical slices, as described in Chapter 5. That approach gives excellent transparency into the present cost and status of an Agile project.

When 75 Percent Truly Means 75 Percent

In an effective Agile team, status reporting is *not* separate from the team's own way of tracking their work—and it's available for all stakeholders to see. Typically, a team board (we've provided a photo of one in Figure 34) shows the stories. The simplest layout will have three states: waiting to be worked on, in progress, and completed.

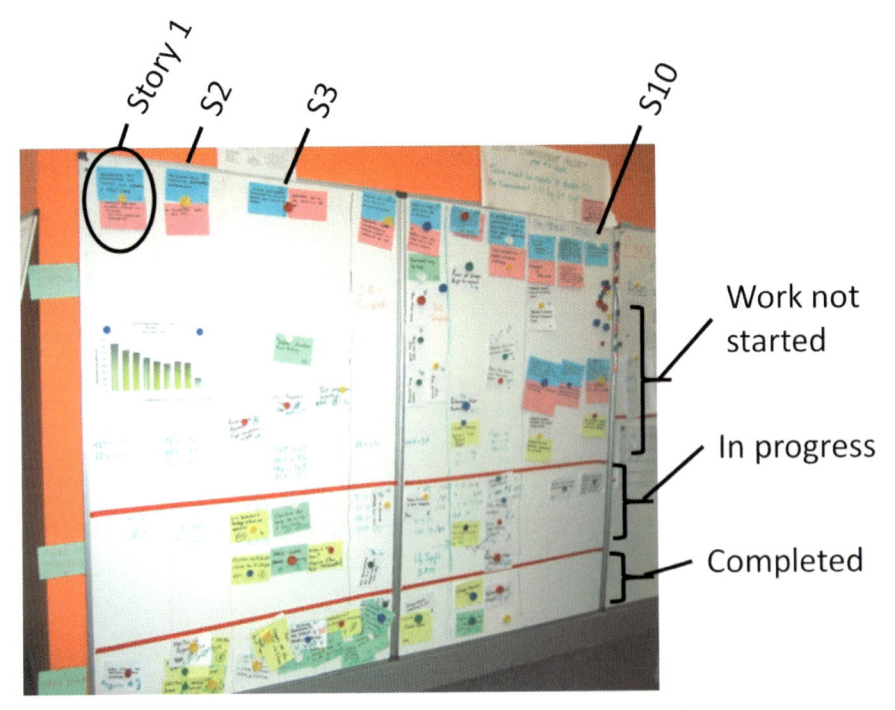

Figure 34 An Agile Team's Story Board

The team story board shown in Figure 34 visually displays all the important facts on the status of this sprint. At the left is the most important story, and it's been completed (all the tasks on white and yellow slips are in the bottom 'Completed' row). So has the next-most important story. The third one hasn't been finished, but the team went on to complete the next two stories. The final three stories have not been started yet. The burndown chart that's hung on the board at the left-hand edge shows that this sprint is nearly over, so the team is having some issues completing the work and may have to de-scope some of it.

A burndown chart shows the work remaining, plotted for some time period. The work remaining may be shown in "story points" or hours, and the time frame is usually the sprint, but can be longer.

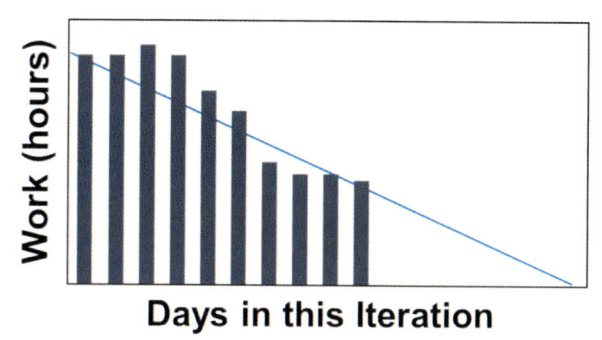

Figure 35 Sprint Burndown Chart: work remaining (hours) each day of sprint

Each day, the team will have a short, 15-minute synchronization meeting (the stand-up). One of the tasks in that meeting is estimating the effort remaining for each task on the plan, as described in Chapter 4. All remaining effort is summed, and that total is today's data point on the burndown chart. If the bars on the chart form a downward trend, the project is going according to the team's estimates. If the trend starts to flatten out, team and stakeholders alike will see that difficulties, either within the team or from outside, have interfered with the progress. You can expect this; the burndown chart allows managers to adjust so as to keep the project on track.

Information Radiators—Easing Communication

The team's story board and burndown chart are examples of "information radiators." While a story board is mainly a mechanism for team members to let each other see where the work stands, managers or other stakeholders can also immediately see all this information without a meeting or even a conversation—just a walk by the team room or other space where everyone posts information they use to signal each other. Similarly, a team member coming back from a day away can immediately see which tasks are open and ready to be started and can start helping right away by selecting a task that matches their skills. No routine status meetings are needed, because the story board and other artifacts are visible all the time.

Here are a few other information radiators that most Agile teams will benefit from having (but you can choose to add more):

- Blockages list
- Risks and issues list
- Team norms poster

A blockages list shows what things outside the team's control are actively impeding their progress. It's updated at the daily synch meeting. In Figure 34, where the third story is not finished, we would expect to see a notation on the Blockage list, because the team should finish that story before moving on to the next one, which is lower priority.

The Risks and Issues list shows those things that are likely to become blockages. The most likely ones are listed first.

A Team norms chart lists the "working agreements" that the team has agreed to live by.

Don't Require Double Reporting

Remember that one of the purposes of all these Agile practices and tools is to make status transparent and remove the need for status meetings. A common and serious mistake is to require the Agile pilot team to report their progress both through Agile means and through traditional progress reports. This combined requirement not only burdens the team with extra overhead, but also communicates that the company doesn't trust that the Agile approach will give real results. This is a time when the company can show support for Agile by taking the burden of status reporting off teams and instead use the Agile information radiator mechanisms.

While we need to measure what the team accomplishes so we have an objective way to know if they're on track, it's important to remember that stakeholders outside the Agile team must avoid using outputs metrics to gauge the team (sprint burndown, velocity, release burndown, or percent utilization of people). These metrics are for the team to monitor their own work. Stakeholders must instead pay attention to outcomes metrics, as described at the end of the previous chapter.

When Things Go Wrong

The information radiators described above are also very useful tools for recognizing and fixing problems.

The story board shown in Figure 34 indicates that the team is not going to be able to finish all the expected work in their sprint. First, note that everyone can see the situation developing; there is no period where people are debating whether to disclose the bad news. There is no "green-shifting" of project status as it travels up the hierarchy. Second, this transparency makes it imperative that something be done while there are still some options open.

In Figure 36, at left no one takes action and the sprint closes with much of the remaining work incomplete. At the right, some scope is cut, enabling the team to cleanly wrap up several of the stories that might otherwise have been left only partially done. You could ask why that matters, since some work is left undone in each case.

Figure 36 Two possible fates for the same sprint

It matters greatly how the incomplete work is distributed. If five percent of each story is not finished at the sprint's end, then nothing can be delivered. If 100 percent of one story—the least important one—is unfinished, then most of that sprint's functionality can be delivered, and the team will learn from whatever forced them to de-scope that story. That's a better scenario. Adjusting scope is the main tool an Agile team uses to stay on track. A story that's been de-scoped may be scheduled for the next sprint. It may no longer be wanted, so could be dropped completely. The point is that everyone takes a fresh look at the project at the end of each iteration. It's time to "inspect and adapt." Use the Iteration review to get a clear view of what's been completed and what now becomes possible: the "inspect" part. Modify what's in the backlog and revisit what to consider for next iteration: the "adapt" part.

Never Add Extra Days onto an Iteration

(Nancy) Iterations (or sprints) should be a fixed length and should not be changed for any reason. It helps the team and everyone outside it to develop a discipline for how they tie all their work together. It can be hard to stick with. I'll explain with a story about a team I coached some years ago.

The team had been using Agile practices for over six months and was seeing the expected improvements in quality and completion of work. They had an iteration that was almost done, but there was some sort of hang-up with the test environment, and it meant that they were going to have to de-scope a few stories if we ended the iteration as planned.

Andy, the team's Agile coach, and Sanjeev, the technical lead, came and reminded me how I always say Agile is for thinking people and you can create new practices as you learn, etc. So they wanted to break the iteration time-box just this once. The stories were really important, and it seemed arbitrary to not deliver them just because of a snag.

"How many days would you add to the iteration?" I asked. They said two, no three. They were certain three would be enough. I persisted. "Suppose some new problem appeared and you couldn't finish it in the three days? Will you then decide to slip it three additional days?" They looked uncomfortable. No way to be sure there wouldn't be a new problem. They decided they would not slip it a second time. I pointed out that they could preserve the credibility that they'd spent months building with their stakeholders by simply not taking that first slip of the initial three days. When they looked at it that way, they could see that they had built trust they didn't want to jeopardize.

Lean Flow

Suppose you have "project A" with 100 tasks, all partly done. You estimate they are on average 60 percent complete. You have a second one, "project B," with 100 tasks of roughly equal size. Sixty of those are done and tested, with the other forty not yet started. How confident are you that project A is really 60 percent done? How about project B?

Of course, it's much clearer that the second project is really 60 percent done—if the finished tasks truly have been tested well. This is exactly the kind of visible, clear signal that an Agile story board should communicate.

Back in the story board in Figure 34, the team has ten stories on the board. It looks like three are not started, three are in progress, and four are done. Wouldn't the project status be clearer if only one story was in progress and all the rest were fully completed or not started? That concept is called keeping your work-in-progress (WIP) small. It's a very simple idea that drives a lot of helpful behaviors. Teams using Lean processes actually set WIP limits on their boards, which forces them to complete older work before opening up a new task. This helps the work to flow faster for Lean teams.

Another closely related idea says that items moving across the story board should not vary too greatly in size. Think about the "project A" and "project B" examples. If all those tasks are exactly equal size, then project B's status is even clearer. Project A's isn't, because the percent done level of its tasks is theoretical.

> "In theory, there is no difference between theory and practice. But, in practice, there is." - Jan van de Snepscheut

In practice, it turns out that if the sizes of stories are kept within a one-to-ten range, that's close enough to have the transparency needed. It's the reason that when estimating user stories, stories larger than ten points should be broken into smaller stories.

Theory and Practice of Sprints

Certain issues come up again and again when teams try to execute their sprints, despite careful planning. Figure 37 shows a chart that a team used in their retrospective. They had a discussion about each of the areas where the sprint burndown chart behaved outside their expectations.

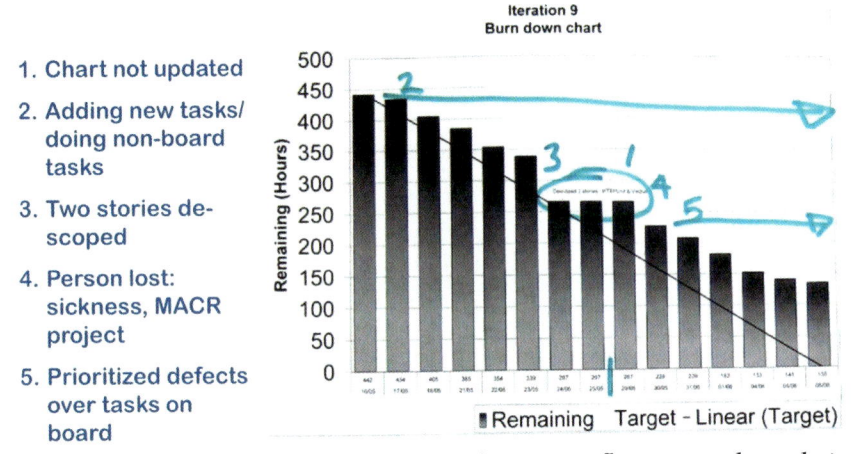

1. Chart not updated
2. Adding new tasks/ doing non-board tasks
3. Two stories de-scoped
4. Person lost: sickness, MACR project
5. Prioritized defects over tasks on board

Figure 37 Notations made by an Agile team reflecting on how their ninth sprint went

Flat areas of the chart at points 1 and 2 happened because they forgot to update the board when they completed tasks, and then they discovered new tasks that should be in the plan after the sprint planning was finished. These happen to every team. Also, they were doing some work that was not even on their story board! That also happens to everyone. They may get asked by another manager to help with something for "just a few minutes."

At point 3, there is a bigger than usual drop, meaning work was either completed all of a sudden or was de-scoped. In fact, they had decided with their product owner to de-scope two stories.

At point 4, we are again in a flat period. Either no work is being done, or new tasks are showing up equal to those being completed. In this case, two team members were away, one sick and one drafted onto another project. This also happens all too often!

At point 5 they shifted their attention away from working the planned stories to troubleshooting defects.

These are the kinds of mistakes that you cannot avoid merely by training teams not to do them. They will happen because they are a sign that a new understanding is emerging at different rates, and in different ways. It's a sign that the team and the organization supporting that team need to pause and ask what is happening and they need to re-align their actions with the goals they've set. In the example shown this team is using their retrospective to do just that.

Creating Balanced Sprints

If you find that the testers in your Agile team are idle at the start of sprints and the developers are idle at the end of them (and testers overworked), you're probably treating each sprint like a miniature traditional project, or waterfall. Some adapt to this by staggering the test work to a parallel work stream. An example would be that while the main team is working on sprint 8, the testers are focusing on the stories that were implemented in sprint 7, and so on, with the testers always staggered one sprint behind. This is not a good move. The reason is that by splitting the work stream, whenever a tester finds an issue that needs a developer's attention, that developer has to switch attention from the current sprint to work that was done weeks ago. This sort of mental context switching has big waste overhead. One reason we want to have teams co-located and 100 percent dedicated to a project is to minimize the mental overhead of attention switching. (See further info in chapter 6 about multitasking.)

Here is a better way to handle the problem of some team members being idle or overworked at sprint boundaries. After your iteration planning meeting is completed, you should immediately start working on the stories, starting with the highest-priority one. As developers work on detailed design for it, the testers can be devising test cases from the story's conditions of satisfaction, prepping data to test with, and so on. The goal is to have the first story completed and tested before starting the second one. As a practical matter, teams will start the second story before the first is fully done so that people aren't idle. If more than two stories are in progress you should ask why. You will probably find that impediments which are allowed to

Agile Methods for Safety-Critical Systems

remain in place are what causes you to keep starting new work instead of finishing the work you've already got open. Rather than do this, recognize that the organization needs to do a better job supporting the Agile team. Have a discussion about how to ensure that impediments are removed the same day they are identified.

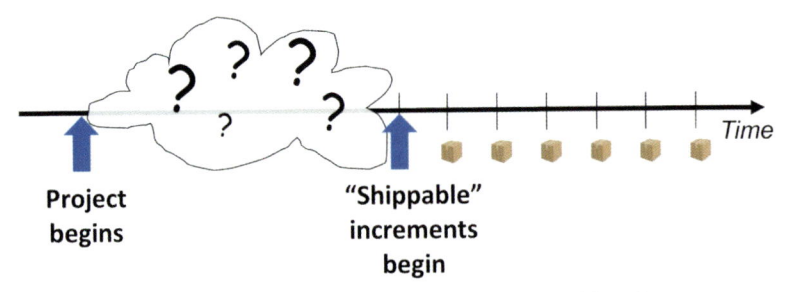

Figure 38 "Iteration Zero" for an Embedded System

Iteration Zero in Embedded Systems

For medical instrumentation and similar h/w s/w products, it may take one or more of "R&D" iterations before you can aim for a truly shippable product increment. For a software-only product, it's common to have just one such iteration, often called Iteration Zero or sprint zero. It is used to get tools set up and a test environment in place—essentially any preparatory activities that are necessary but do not produce software.

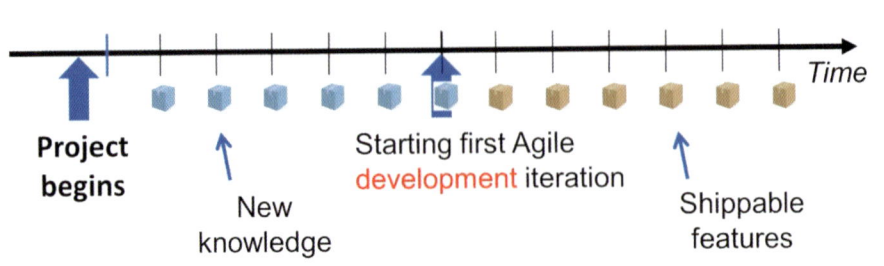

Figure 39 Building Knowledge in Early Iterations for an Embedded System

In embedded systems work, it may take more than just a few weeks (the length of one sprint) to be ready to have potentially shippable product increments at the close of each sprint.

Before an initial version of a medical instrument can be tried in a realistic setting, there are software elements that need to be built (drivers, boot code, interfaces to displays, etc.), and hardware elements (circuit boards, cabling, sensors…) that can take months to complete. That's long enough for an Agile team to lose its way if our only measure of progress is shippable product increments.

When a product is an embedded system, as with many medical devices, the early stories tend to come from inside the team rather than from end customers, because they are exploring feasibility of the product.

When early stories come from implementers, not business-side stakeholders, it can be an education for the Product Owners. They gain insight into how much work is below the water line in the technical layers. It may be best for the Product Owner to have a technical background or to work closely with someone who does.

Be careful that you don't stay in this mode too long; team members should not be the source of most stories after the initial few iterations. Usually you can implement some user-facing stories even very early—try to do this.

Early iterations serve "near" customers…

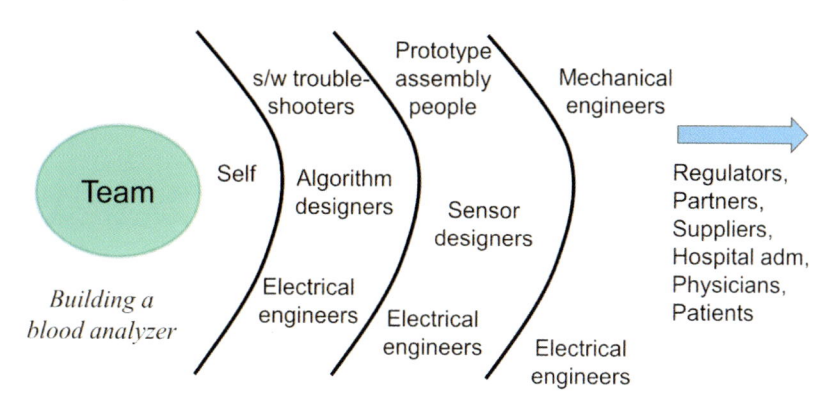

Team	Self	s/w trouble-shooters	Prototype assembly people	Mechanical engineers	Regulators, Partners, Suppliers, Hospital adm, Physicians, Patients
Building a blood analyzer		Algorithm designers	Sensor designers		
		Electrical engineers	Electrical engineers	Electrical engineers	

Figure 40 Targets for Early Stories in New Product Development

Here are some sample early-stage stories' headlines:

- Send 'Hello world' to display unit via CAN bus
- Create routine to verify all CCD elements as part of POST (Power-On System Test)
- Create routine to verify all on-chip RAM is working; use cyclic redundancy check with 8-bit checksum
- Set up boot-loader address table [test it manually—needed only for a short time, so can skip making an automated test]
- Read raw data from sensor hardware and store in non-volatile memory (to pass to algorithm designers)
- Read test point voltages and report to electrical engineers via static table; use D/A counts

Strategic release planning across many iterations also needs to be done, and that entails collaboration between software, hardware, and business:

- Decide which customers we're serving when
- Envision minimum marketable feature (MMF) sets
- Interleave h/w and s/w revisions
- Decide which versions will not be backward compatible with older hardware
- Plan replacement/upgrade of prototype hardware
- Plan data development strategies (e.g., characterization of blood types, calibrations)

Accelerating Learning Through Deep Focus and Other Agile Practices

A dedicated team is crucial to tracking—and making—progress in Agile development. When a company launches an Agile pilot project, they almost inevitably make the mistake of allowing other projects to pull people off the Agile team or to swap people between the Agile team and other teams. Traditional teams can tolerate this, but Agile teams cannot. Agile's practices are heavily dependent on frequent communication between team members and on the fluency they gain in doing the work in the vertical slices manner discussed earlier. When team composition changes suddenly, it is a very big disruption.

Agile team members get into an extremely focused way of working throughout the day. It is this strong focus that has a bearing on safety-critical work; it amounts to there being fewer opportunities for things to slip through the cracks. Practices like pairing and whole-team programming accelerate work as well as providing ongoing review as the work is being done.

Pairing (or pair programming) can be used effectively for requirements, design, and testing as well as coding. There are several benefits to this method:

- Better designs
- Effective training/mentoring
- Reduced risk of knowledge loss
- Improved quality via constant review

Research has shown that implementing pair programming confers a "48 percent increase in correctness for complex systems, but no significant difference in time, whilst simple systems had 20 percent decrease in time, but no significant difference in correctness."[49] While the increase in speed for simple tasks is nice, the significant thing here is that the complex work—which is the breeding ground for costly defects—is more accurate while still getting completed at the same rate as two people working separately.

Decreasing defects in complex work not only saves costs but also has a direct impact on risk. Pairing can serve as peer review if the company addresses regulatory concerns:

- Established acceptance criteria
- Defined reviewer qualifications
- Documentation of results

There is no regulatory objection to this practice. AAMI TIR45 says *"Pairing as a form of peer review (as part of the overall verification process) must satisfy the same requirements as any other method of peer review, addressing the considerations described [in TIR 45]."*[33]

Learning can also be accelerated through a practice called Whole-Team Programming. Starting in 2011, this coding practice based on pairing grew out of daily learning sessions being used by a software development team in San Diego. They wanted to learn more about TDD and refactoring to improve their present codebase. They began using group learning-plus-coding activities for dealing with a difficult project where the knowledge of the whole group, activated together, was a real help. The effect was so strong and positive that they never went back to solo coding. They began working together, as a group of six or fewer, all day every day.[50]

Soon they gave it the name "Mob Programming," and other software teams became curious and began to try it. This technique has been rapidly spreading all over the world. There is nothing in the technique that is objectionable to regulatory bodies, since they do not take a position on what development process a company uses.

Chapter 11 – Scaling Up

Friction is Common

Having one or a few teams that work faster and more efficiently than the rest of the organization can easily cause strain—just like bolting a jet engine on top of your Pinto.

Compared to traditional teams, Agile teams put out more frequent releases, need more stakeholder decisions, need broader access to test equipment, and need different supervision. You need a guidance system: an Agile Adoption project to head off the inevitable friction.

Agile Adoption Project

The Agile Adoption project must aim to charter, support, and learn from the individual Agile development projects. The issues faced by Agile development projects will fall into patterns, which show where the organization's support for an Agile way of working is weak. Before those issues occur, you can anticipate and mitigate many common problems.

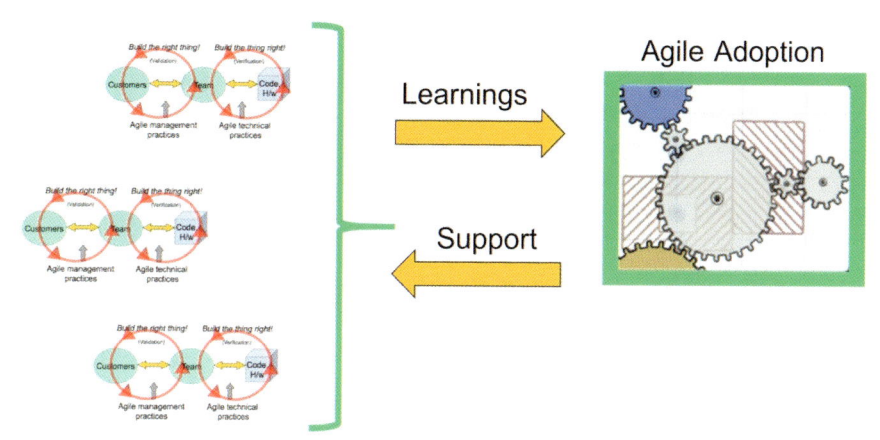

Figure 41 The Agile Adoption Project

All roles in the lean/Agile organization need the support to carry out their tasks: the product sponsor, the product owner, the development team, and the coach/Scrum Master. In many organizations, an Agile "champion" helps ensure that those roles are properly supported and protected.

The Agile Adoption project needs to gather learnings from experiences of the teams and provide support so those teams can work effectively without interference.

Start Small

Moving Agile to a larger scale than one or a few teams is a concern for many companies, since their projects are either spread across multiple groups or carried out in multiple locations. Successful examples exist, and several methodologies have been proposed.

However, the crucial step before worrying about larger scale is to get the Agile approach working at small scale. You should not consider expanding your Agile program until you have the dynamics running smoothly in your pilot team or small group of pilot teams. Where small-scale teams have succeeded and management is committed to making the new approach work, you have the best possible chance to figure out how to scale up Agile.

Here are some classic signs that you need to address issues before attempting to expand your Agile program to more teams:

- The backlog has run out of stories that are ready for scheduling in a sprint by the time you need to start the next sprint (ready means the conditions of satisfaction are agreed, and the team has discussed the story and has given it an estimate)
- Your team is unable to build and test the set of stories within the same sprint
- Regression test of prior stories is taking more and more of the team's energy each sprint
- Your team is not routinely completing all the stories they plan to deliver in each sprint (stories should only occasionally be de-scoped)
- The number of defects being produced in new code is not decreasing

A mistake that is all too common is believing that the problems you're now having (missed commitments, still too many defects, too little customer feedback...) will be alleviated by scaling Agile to many more teams.

The following case study shows how important it is to have executive sponsorship and real understanding of Agile, if your Agile Adoption program is to succeed.

Case Study: Agile Adoption Can Fail Even When All Teams Succeed!

Summary

Property Holders Insurance, a company with over 20,000 employees, launched an ambitious Agile adoption program, with top expert help, after an Agile pilot project delivered its full scope in less than half the estimated time with less than half the usual quality issues. Dozens of Agile teams were created, and they delivered. But although the business desperately needed faster time-to-market performance from IT, within two years the Agile adoption program was completely scrapped.[51, 52]

What happened -

Ben, a project manager at PHI, had learned enough on his own about the Agile framework—Scrum—to give it a try. His project was so successful (thirteen months' worth of work delivered in six, and with quality about twice as high as before) that the Director of Product Delivery took notice. In time, Ben was put in charge of expanding the process to other teams. Note that the highest-level sponsorship of Agile was with Ben, a project manager. Although the Director of Product Delivery took notice and liked what he saw, his understanding of Agile was too superficial for us to say sponsorship existed with him.

Eventually Ben was put in charge of expanding Agile. His goal was to keep iterations short (two to four weeks) and deliver as many of them as possible all the way through production. He realized that he would need to expand the number of Agile teams quickly, because executive management had shifted to demanding much faster project timelines from the Project Delivery organization. Aggressive top-level targets for the whole company were set for July, just five months away.

By May, Ben had engaged thirteen external Agile coaches and was set to rollout forty new Agile teams. The coaches had tailored Agile training as a two-day course for PHI, and this was launched to 500 employees. "My biggest mistake was not going slower in the Agile rollout," Ben said in hindsight. But the pressure from the executive management level for July deliverables was tremendous, and there was no way he could push back without that being viewed very negatively. It soon became apparent that these Agile coaches had different interpretations of what should be allowed, and their varied experience levels began to show. Some coaches told teams that every single story must carry customer value, while others said it's okay to have infrastructure stories. These and other instances of variation served to confuse employees.

"We hadn't defined our Agile methodology clearly, and this was now hurting us," Ben said. With so many new teams starting and with thirteen different external coaches, there was no way Ben could oversee what they were all doing.

The teams were able to govern their iterations reasonably well. They tracked their productivity and used past performance to negotiate the content of iterations. They worked very hard to keep the commitments they made. But there was contention for resources, and people were sometimes pulled from Agile teams to be used for urgent work elsewhere.

Each Agile team was made up of a Product Manager as team leader, developers, testers, technical lead, and analysts. The Product Manager was acting in the Scrum Product Owner role. Software releases were never seen by actual customers until the whole project was completed. This is a compromise in Scrum's methods. Scrum also recommends that teams should be using practices like test-driven development, continuous integration, and automated unit tests, and these new teams were varying greatly in their use of such practices. They hadn't had time to learn those skills in the abbreviated training. As a result, quality suffered. They were producing defects at a much higher rate than Ben's pilot Scrum team.

The Agile teams generally achieved their July deliverables, but quality had suffered. They'd worked long hours and made sacrifices to get the releases out, and then were told that it wasn't enough—more was needed. Morale took a nosedive.

A power struggle developed between Adrian, the VP of IT, and John, the Senior VP of Product Delivery, where Ben worked. Against Adrian's criticisms of product quality and too much time wasted on elaborate up-front analysis, John responded "You can't do projects without analysis, design, and construction. *We* own that." Adrian's view was that Agile is only concerned with the production of software—not product management—and so IT should own the Agile sponsorship. Ben took this issue all the way to the CEO, confident that he'd agree Agile sponsorship should stay with John's organization.

The CEO backed Adrian, and with that, the ownership of Agile adoption moved to the IT organization.

The rest unwound quickly. Ben brought in a new senior coach, whose work was impeded when Adrian's group would not implement the data collection needed for Agile and Lean metrics. The Agile program was stalled.

That December, a new CEO was named; a change that many said was mainly due to the alarming drop in morale among those who'd been part of the massive Agile rollout. The new CEO's message was that we will have steady predictable progress from now on. No surprises. One of the criticisms of Agile had been that although time to market was great, predictability was poor. Finishing a project in less than the projected time was viewed as a lack of predictability. The mass of teams started in May would not have reached steady state by July. The earlier teams were pilots. Expecting predictability by that definition was simply unrealistic. Aside from the early stage of the teams, shortcutting the agile technical practices meant that defects in the code would continue to generate unpredictable amounts of rework.

Within a couple more months, the new CEO had quietly allowed the number of Agile teams to dwindle. By February there were no more of them.

Four Pillars of Agile Adoption

If a business is to sustain and spread the success of an Agile pilot project throughout the enterprise, there are actually four change initiatives that must be managed successfully. Even more critical, they must occur *simultaneously*. These pillars of Agile Adoption are below:

- Teams must be able to produce defect-free software sustainably
- Teams must consist of empowered, engaged people

- Workflow to the Agile teams must be controlled via a "pull" system
- Lean portfolio management must be used to control workflow for the organization

Unless *all four* of these change initiatives are running successfully, characteristic problems arise, as seen in the case study.[53]

Where Lean Meets Agile

The first three pillars of Agile adoption are covered by the popular Agile methodologies, but this issue of governing the work stream at the organizational level takes us over to Lean territory. Unless a lean discipline is used to decide what work an organization undertakes, it runs the risk of spreading its energy too thinly. Weakly supported projects will thrash and waste resources. A company that regularly completes forty to fifty projects a year should not have 500 active projects in its portfolio!

Agile Habitat

Another impediment to scaling up can be resources. One Agile pilot project can be sustained by almost any company. It will place more or less demand on various other departments and infrastructure, but it can be coped with. As the number of Agile teams increases, pressure on certain resources (testing environments, Product Owner attention, team rooms, etc.) reaches a point where something has to give. Either the Agile teams will be reined in and forced to make more and more compromises in their roles and practices, or else managers will recognize that they have to make the organization into a better habitat for Agile teams.

A perfectly skilled and competent Agile team may still fail. Once Agile has been established and the company's gained trust in this new way of working, the highest-priority projects should go to Agile teams. Portfolio managers will have to regularly kill off those projects that cannot deliver or that are vying with Agile projects for resources. If the project portfolio isn't managed according to lean principles, even fully competent Agile teams will be spread too thin to succeed.

Takeaway Lesson

If teams have mastered defect-free code, are empowered, and their work stream is properly controlled, they can still be destroyed by the failure of managers to match the *organization's* workflow to its sustainable capacity. One more time: Managers *at every level* have got to buy into the idea that they must never jam more work into a pipeline than its *proven* capacity.

> "The art of progress is to preserve order amid change and to preserve change amid order." – Alfred North Whitehead

Agile at the Enterprise Level

Because this book is purely a primer, we have elected not to discuss the merits and drawbacks of various models for Agile at scale. The effects we've described in this Scaling Up chapter can appear quite early in your move to Agile, long before you're anywhere close to enterprise scale. It's important to be alert to reading the signs of strain between the Agile teams and the surrounding organization, because they are the most reliable guide for what steps you should take next.

There is a tremendous urge these days to unveil a complete model for Agile organizational change and then to prescribe the steps for making it happen. While there are recurring patterns, there simply is no model complete enough to justify going so far to a top-down approach.

(Nancy) Some years ago I was doing genealogical research that taught me a lesson I found useful in the work of Agile change-making. It started when I read about an American Civil War general who was from our area. I wanted to know if my family is related to him. So I tried to start by looking up the family surnames associated with him. I wasn't getting anywhere; there were so many possible directions to go in, and no clue which might be most promising. Find a common ancestor? Look for descendants who might have married someone in our family? I read a book on how to do that kind of research, and the author said many people start wrong by picking some famous person or king that they hope to be descended from and then trying to work down from there to themselves. Instead, the book said to start with what you know for certain, and what your parents and grandparents can tell you. That way, all the effort you put in is relevant to the people who are on the path you want to understand. When you start with the famous person, all your effort is possibly a waste. You might never make a connection to your own family. The key advice: Always work from the known, moving into the unknown.

I immediately switched my strategy, and very soon I found the info on all of the family back through the middle 1800s. I was also able to eliminate the possibility that we were descended from that general. In the process, I found many more paths to explore that were all 100 percent relevant, because I was working from the known into the unknown.

For Agile change within companies, if you read the signs that are there—the blockages teams experience, the recurring issues they visit in retrospectives, the difficulties that supporting groups (IT, marketing, HR, quality assurance, regulatory affairs, field service) express with respect to Agile teams—you'll have a set of known facts on the ground to work from. Get good at reading the signs. Learn all you can about Agile and Lean basics and about others' experience trying to change things. Then you will be able to see a viable path to pursue in helping your company become more Agile.

Chapter 12 – Simple, But Not Easy

Don't Let the Newborn Chick Get Eaten!

In any organization there are people, career paths, and processes that resist innovation. These organizational antibodies can be very strong, even in companies where a history of failures shows that they desperately need a different approach. The Harvard Business Review article mentioned in Chapter 3[35] points out that "More than 70 percent of Agile teams report tension between the way the team operates and the way the whole organization operates." Leadership and protection are absolutely vital to pursuing an Agile transformation, even for just one team/project.

Once your Agile pilot team gets its work rolling, you will need to shepherd the change process—to inoculate the organization against those organizational antibodies. Not all the company can change at once, and those outside the Agile group may want to snipe as early stumbles occur. You will need to patrol this border between Agile and traditional.

Fixed vs. Growth Mindset

Regardless of the approach you take to implement Agile in your company, the important thing to remember is that Agile is a mindset. Software development has many inherent uncertainties; a Fixed Mindset seeks to reduce uncertainty by nailing everything down, while a Growth Mindset seeks to reduce uncertainty by discovering and learning, and is inevitably much more productive.[54]

Reverting Under Pressure

(Nancy) It's natural to stick with what you know well when the pressure is on. If you're on your way to a job interview, it's not a time to try driving a new shortcut you just heard about. Better to do that when you don't have a deadline.

This same effect applies when a team is new to Agile practices. When the pressure is high, and people are starting their first sprint, they will feel a very strong pull to go back to what they know best—waterfall style development! This is only natural.

(Nancy) You can add support for the new learning, and you can look for ways to reduce the pressure. Sometimes in a team's first sprint, I'll coach them and management to view it as a calibration sprint, meaning that they don't have to promise just how many stories they will complete. Instead, they'll just size the stories and finish them one by one to discover how many they can complete. It gives them an initial reading on their team capacity, and that makes it easier for them to estimate next time. This is a great way to reduce pressure, because the team and their managers are onboard with it.

If you're a peer or manager of people learning Agile, don't get upset with them if they backslide. Look to see if the pressure is just too high, or if they don't have the amount of support (info, tools, coaching) that they need. Ask yourself what you can do to alleviate the pressure or boost the support.

Simple But Not Easy

The simplicity of Agile can be deceptive. We expect complex things to be hard – like running a business or learning string theory. We expect simple things to be easy; it's logical. It's interesting that some complex things are easy too. Examples include walking and recognizing faces, but it turns out that we have special biological wiring for things like that so they seem easy. Agile falls in the opposite combo – like hitting a baseball it is conceptually simple but takes considerable practice to do well. We compared it at the beginning of this book with learning a language or learning to play a musical instrument; there is some "book learning" but real understanding comes only with a great deal of practice.

Agile Methods for Safety-Critical Systems

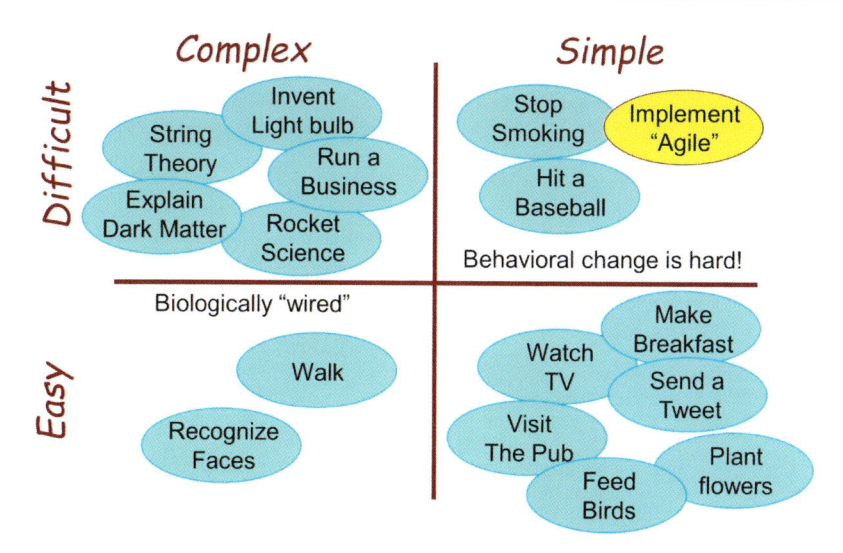

Complex | **Simple**
- Difficult: String Theory, Invent Light bulb, Explain Dark Matter, Run a Business, Rocket Science | Stop Smoking, Implement "Agile", Hit a Baseball
- Behavioral change is hard!
- Biologically "wired"
- Easy: Walk, Recognize Faces | Watch TV, Visit The Pub, Make Breakfast, Send a Tweet, Feed Birds, Plant flowers

Figure 42 Easy or Difficult is not the same as Complex vs. Simple

We've aimed to show how the lean and Agile concepts are both flexible and strong: they reach beyond development. They support more than just development of software and hardware by covering product definition through product release, and they address planning from the iteration level all the way up to the portfolio level.

Vital to gaining the flexibility and strength is the job of bridging your silos. No longer can marketing, clinical, field service, training, engineering, and manufacturing lob issues over the wall at each other.

From the quality and regulatory point of view, keep these facts in mind:

(a) No standard or regulatory body requires a specific development lifecycle
(b) The standards do require us to analyze safety risks iteratively, which fits well with the Agile approach
(c) Quality regulations (21 CFR Part 820) and standards (ISO 13485) require us to document our activities, but they do not dictate any lockstep order of those activities.

Making the transition to Agile may sound simple, but not everything that is simple is necessarily easy. Changing entrenched company culture is hard. Engaging an experienced coach is your best investment if you're truly looking for the advantages Agile can bring you.

Afterword: Recommended Reading

If you've gotten this far, undoubtedly you're curious and want to find out more. The references we've listed below will help fill in the information we didn't have space to cover in our primer.

Agile and Lean Methodologies, including job roles

These give a good understanding of the core ideas in Agile and Lean, and how they originated. The last two are more focused on current practices in Scrum.

Lean Software Development, Mary and Tom Poppendieck

Agile & Iterative Development: A Manager's Guide, Craig Larman

Mob Programming: A Whole Team Approach, Woody Zuill and Kevin Meadows

Kanban From the Inside, Mike Burrows

Lean-Agile Pocket Guide for Scrum Teams, Alan Shalloway and James Trott

The Scrum Guide, Ken Schwaber and Jeff Sutherland, available at **http://www.scrumguides.org/download.html**

Technical References (Medical Devices)

These are technical references with valuable information beyond the standards mentioned in the text. They are copyrighted documents, and may not be readily available to people not currently working in medical device companies.

AAMI TIR45: 2012: Guidance on the use of AGILE practices in the development of medical device software, Association for the Advancement of Medical Instrumentation, 20-August-2012.

ANSI/AAMI/IEC TIR80002-1:2009, Medical Device Software - Part 1: Guidance on the application of ISO 14971 to medical device software, Association for the Advancement of Medical Instrumentation, 3-September-2009. (a particularly helpful guide on how to approach risk management for medical device software - incorporates the text of ISO 14971, with comments after each section as to how it applies to software)

Agile Technical Practices for Software

Methodologies have to be expressed in actual practices - and these are great starting places to understand software development and test as they should be done in modern Agile companies. Although many claim they are agile, we see only a fraction of them actually using the practices.

Agile Testing, by Lisa Crispin and Janet Gregory

Test Driven Development for Embedded C, by James Grenning

Refactoring: Improving the Design of Existing Code, by Martin Fowler

Clean Code, by Robert C. Martin

The Art of Agile Development, by James Shore and Shane Warden

Continuous Delivery, by Jez Humble and David Farley

Teamwork

Teamwork in Agile companies is different - more collaborative, and more self-directed. More autonomy demands that you be more responsible as a group. That requires significantly deeper skills in facilitation, negotiation, and simply patience in dealing with each other. Yes, the Mob Programming book is repeated here because good teamwork is its chief foundation.

Getting to Yes, by Roger Fisher and William Ury

Teaming, by Amy Edmondson

Mob Programming: A Whole Team Approach, Woody Zuill and Kevin Meadows

Agile Retrospectives, by Esther Derby and Diana Larsen

Agile Coaching, Rachel Davies and Liz Sedley

Decide how to decide, (article) Ellen Gottesdiener - available here **https://www.ebgconsulting.com/Pubs/Articles/DecideHowToDecide-Gottesdiener.pdf** and also in the form of a blog here **https://www.ebgconsulting.com/blog/decide-how-to-decide-empowering-product-ownership/**

Requirements / Vision

Moving from specifications up-front to a more interactive way to keep control over projects is a major step in moving to any kind of

Agile methodology. These are good starting points to deepen your understanding of that path.

Discover to Deliver: Agile Product Planning and Analysis, Ellen Gottesdiener and Mary Gorman

Agile Estimating and Planning, Mike Cohn

User Story Mapping, Jeff Patton

Impact Mapping, Gojko Adzic

Agile Management

Executive management has a vital role to play, not only in guiding the shift to Agile, but also for leading in a different way that leverages the new reality that the deepest know-how in high tech business (moving to be true of all business!) is at the "leaf nodes" not at the top of a hierarchical pyramid. These books are a great starting point to delve into all the implications of that.

The Age of Agile, Stephen Denning

Leading Lean Software Development, Mary and Tom Poppendieck

Succeeding with Scrum, Mike Cohn

Wiki Management, Rod Collins

Radical Project Management, Rob Thomsett

Joy, Inc., Rich Sheridan

About the Authors

Nancy Van Schooenderwoert is President and Principal Coach at Lean-Agile Partners. Nancy was among the first to apply Agile methods to embedded systems development, as an engineer, manager, and consultant. She has led Agile change initiatives beyond software development in safety-critical, highly regulated industries, and teaches modern Agile approaches like Mob Programming, Agile Hardware, and Lean development methods.

Initially working as an electronics designer and software engineer in flight simulation, she later focused on software engineering. Nancy has worked coaching Agile teams in the USA, UK, and Germany. Her coaching extended to their work with their teams in Japan, India, China and other countries.

Nancy's experience spans embedded software and hardware development for applications in aerospace, factory automation, medical devices, defense systems, and financial services. Her coaching practice spans delivery teams to middle and upper managers. She is a regular presenter at Agile-related conferences worldwide. She is a founder and past president of Greater Boston's premier Agile user group, Agile New England.

Brian Shoemaker consults for healthcare products companies on computer system validation, software quality assurance, and electronic records and signatures. He has conducted validation both on product software and on internal software, developed software quality systems, audited software quality processes (including Agile methodology), and evaluated 21 CFR Part 11 compliance.

Brian's strengths include:
- Application of software development lifecycle and quality systems in FDA-regulated environment.
- Direct experience applying Electronic Records and Signatures rule (21 CFR Part 11)
- Multidisciplinary team leadership: coordinating work, obtaining consensus, managing by influence
- Applied knowledge of quality system components, SOP design, and auditing.

Much of Brian's activity has been in validation and software quality for medical devices, but some of his projects have been for companies in clinical trial data managements. His clients have been in fields as diverse as medical device engineering, medical imaging, medical-device fabrics manufacturing, contract lyophilization, clinical trial software, dental prosthetics, and bone-repair implants. He has worked with companies in Germany and Switzerland as well as the U.S.

Nancy and Brian provide specialized consulting to medical device companies, covering all the latest regulatory developments in the US and EU, plus the Agile practices for documentation and incremental risk assessment that is essential to the medical device industry.

Their course "Introduction to Agile Adoption for Regulated Medical Software" has proven popular with managers and senior technical leaders in medical device companies. For more information please see **www.AgileMethodsforSafetyCriticalSystems.com**.

References

[1] Leveson, N.G., *Safeware - System safety and Computers. 1 ed.* 1995: Addison-Wesley Publishing Company Inc. [Revised version at: http://sunnyday.mit.edu/papers/therac.pdf]

[2] The FDA provides a searchable database of medical device recalls at https://www.accessdata.fda.gov/scripts/cdrh/cfdocs/cfRES/res.cfm. Recalls are cited here by their recall number.

[3] FDA Recall Number Z-1021-04

[4] FDA Recall Number Z-1334-04

[5] FDA Recall Number Z-1545-05

[6] FDA Recall Number Z-0495-2018

[7] FDA Recall Number Z-1022-05

[8] FDA Recall Number Z-0342-05

[9] FDA Recall Number Z-0598-2016

[10] FDA Recall Number Z-0625-2018

[11] FDA Recall Number Z-1934-2010

[12] Poppendieck - cook vs. chef concept

[13] International Organization for Standardization, ISO 13485:2016 "Medical devices - Quality management systems - Requirements for regulatory purposes." 1 March 2016.

[14] **http://agilemanifesto.org/**

[15] Sidky, Ahmed, "The Agile Mindset." Available at http://www.softed.com/assets/Uploads/Resources/Agile/The-Agile-Mindset-Ahmed-Sidky.pdf. Reproduced with permission.

[16] H. Alemzadeh, R. K. Iyer, Z. Kalbarczyk, J. Raman, "Analysis of Safety-Critical Computer Failures in Medical Devices." *IEEE Security & Privacy*, July-Aug. 2013.

[17] Version One's ninth State-of-Agile survey, **http://info.versionone.com/state-of-agile-development-survey-ninth.html**

[18] Beck, Kent, and Cynthia Andres, *Extreme Programming Explained: Embrace Change, 2nd Edition*, Boston, Addison-Wesley, 2005.

[19] Conversation with Ward Cunningham, 2004.

[20] Poppendieck, Mary and Tom, *Implementing Lean Software Development*, Addison Wesley, 2007. (see chapter 2)

[21] International Organization for Standardization, ISO 14971:2007 "Medical devices - Application of risk management to medical devices." 1 March 2007.

[22] American National Standards Institute and Association for the Advancement of Medical Instrumentation, ANSI/AAMI/IEC 62304:2006 "Medical Device Software - Software Life cycle Processes." 17 July 2006 (with Amendment 1:2005).

[23] M. Mah, "How Agile Projects Measure Up and What This Means to You." *Cutter IT Journal* vol 9, no. 9, Sep 2008.

[24] N. Van Schooenderwoert, "Embedded Agile Project by the Numbers With Newbies." Agile 2006 conference report.

[25] Capers Jones, "Software Quality in 2002: A Survey of the State of the Art." Presentation to Boston SPIN, Oct 2002.

[26] Fowler, Martin, *Refactoring: Improving the Design of Existing Code*, Reading MA, Addison-Wesley, 1999.

[27] Capers Jones: private communication with Nancy Van Schooenderwoert, referring to dataset "Average Productivity Rates in Function Points per Staff Month by Type and Size of Software Project."

[28] Jenks, J.R. (AgileTek), and R. Rasmussen (Abbott), "Moving to Agile in an FDA Environment: An Experience Report." Presented at Agile 2009, Chicago IL, August 2009.

[29] Source: ABC News in California, **http://abc7news.com/technology/san-mateo-cyber-security-firm-uncovers-malware-on-medical-devices/1757268/**

[30] Dr. Jack Lewin, MD, Presentation: "Healthcare Organizations are Under Attack – High-value Resources are At Risk!" Software Design for Medical Devices conference, Munich, 19-22 February, 2018.

[31] Ponemon Institute, 2017. Medical Device Security: An Industry Under Attack and Unprepared to Defend – as referenced by Hannah Murfet (CQP MCQI, BSc,

DipQ), Presentation: "'Cyber Threat' – Considerations for Risk Management in the Product Lifecycle", Software Design for Medical Devices conference, Munich, 19-22 February, 2018. This Presentation was based on 'Murfet and Warner, 2018; A critical review of risk management approaches for medical device cybersecurity; Cranfield University'.

[32] Hannah Murfet (CQP MCQI, BSc, DipQ), Presentation: "'Cyber Threat' – Considerations for Risk Management in the Product Lifecycle", Software Design for Medical Devices conference, Munich, 19-22 February, 2018. This Presentation was based on 'Murfet and Warner, 2018; A critical review of risk management approaches for medical device cybersecurity; Cranfield University'.

[33] Association for the Advancement of Medical Instrumentation, AAMI TIR45:2012 (Technical Information Report) "Guidance on the use of AGILE practices in the development of medical device software." 20 August 2012.

[34] http://sdlearningconsortium.org/

[35] Rigby, Darrell K., Jeff Sutherland, Hirotaka Takeuchi, The Big Idea: Embracing Agile, *Harvard Business Review*, May 2016, p. 40.

[36] See James Grenning's blog entry "Planning Poker Party (The Companion Games)" at **http://blog.wingman-sw.com/archives/36**

[37] Sheridan, Rich, *Joy, Inc.*, Portfolio Penguin, 2013, pp 73-75.

[38] **https://www.mountaingoatsoftware.com/agile/user-stories**

[39] Patton, Jeff, *User Story Mapping: Discover the Whole Story, Build the Right Product*, Sebastopol CA, O'Reilly Media, 2014.

[40] **https://www.mountaingoatsoftware.com/agile/scrum/resources/overview**

[41] Mamoli, Sandy, *Creating Great Teams - How Self-Selection Lets People Excel*, Pragmatic Bookshelf, 2015.

[42] **http://www.fastcompany.com/articles/2008/07/interview-gloria-mark.html**

[43] Dr. Jordan Grafman, chief of the National Institute of Neurological Disorders and Stroke at the National Institutes of Health, quoted by by Melissa Healy in "We're all multi-tasking, but what's the cost?" *Los Angeles Times*, July 19, 2004.

[44] Weinberg, Gerald M., *Quality Software Management: Systems Thinking*, New York, Dorset House, 1991.

[45] U.S. Food and Drug Administration, General Principles of Software Validation; Final Guidance for Industry and FDA Staff, January 11, 2002.

[46] Stacey, Ralph D., *Strategic Management and Organisational Dynamics: The Challenge of Complexity, 3rd Ed.*, Financial Times, Harlow, England. Note that Dr. Stacey no longer uses this figure because he has found that it suggests managers can decide which kind of situation they are in and then choose the appropriate tools. The real world is never so neat as to make this a practical expectation. However we choose to use the figure for making our point that Agile approaches handle situations that traditional methods do not, and cannot.

[47] Adzic, Gojko, *Impact Mapping: Making a Big Impact with Software Products and Projects*, Woking, Surrey UK, Provoking Thoughts, 2012.

[48] Thomsett, Rob, *Radical Project Management*, Prentice Hall PTR, 2002.

[49] Arisholm, Erik; Hans Gallis, Tore Dybå, Dag I.K. Sjøberg, (February 2007) "Evaluating Pair Programming with Respect to System Complexity and Programmer Expertise." *IEEE Transactions on Software Engineering* 33(2): 65–86. doi:10.1109/TSE.2007.17. Retrieved 2008-07-21.

[50] Zuill, Woody and Kevin Meadows, *Mob Programming: A Whole Team Approach*, Leanpub, **https://leanpub.com/mobprogramming**.

[51] Nancy Van Schooenderwoert, "The Four Pillars of Agile Adoption." *Cutter Executive Report* vol. 9, no. 6 (June, 2008).

[52] The real name of the company is not being used, since some details are sensitive.

[53] The need for these four parallel initiatives came out of a conversation with Mary Poppendieck.

[54] Dweck, Carol, *Mindset: The New Psychology of Success*, Ballantine Books, 2006.

18401235R00072

Printed in Great Britain
by Amazon